T0353439

Scientific Computation

Using real-life applications, this graduate-level textbook introduces different mathematical methods of scientific computation to solve optimization problems through examples ranging from locating an aircraft, to finding the best time to replace a computer, analyzing developments on the stock market, and constructing phylogenetic trees.

The textbook focuses on several methods, including non-linear least squares with confidence analysis, singular value decomposition, best basis, dynamic programming, linear programming, and various optimization procedures. Each chapter solves several realistic problems, introducing the modelling optimization techniques and simulation as required. This allows readers to see how the methods are put to use, making it easier to grasp the basic ideas. There are also worked examples, practical notes, and background materials to help the reader understand the topics covered. Interactive exercises are available at www.cambridge.org/9780521849890.

Gaston H. Gonnet is a Professor in the Department of Informatik at ETH, Zürich. His research interests lie in symbolic and algebraic computation, system development, limit and series computation and heuristic algorithms, computational biochemistry algorithms, and the design and analysis of algorithms.

Ralf Scholl teaches mathematics, physics, and science and technology at the Geschwister-Scholl-Gymnasium in Stuttgart. His interests lie in the didactics of teaching science and mathematics at high school to graduate level.

Scientific Computation

Gaston H. Gonnet

ETH, Zürich

Ralf Scholl

Geschwister-Scholl-Gymnasium, Stuttgart

CAMBRIDGE
UNIVERSITY PRESS

CAMBRIDGE
UNIVERSITY PRESS

Shaftesbury Road, Cambridge CB2 8EA, United Kingdom

One Liberty Plaza, 20th Floor, New York, NY 10006, USA

477 Williamstown Road, Port Melbourne, VIC 3207, Australia

314–321, 3rd Floor, Plot 3, Splendor Forum, Jasola District Centre, New Delhi – 110025, India

103 Penang Road, #05–06/07, Visioncrest Commercial, Singapore 238467

Cambridge University Press is part of Cambridge University Press & Assessment,
a department of the University of Cambridge.

We share the University's mission to contribute to society through the pursuit of
education, learning and research at the highest international levels of excellence.

www.cambridge.org
Information on this title: www.cambridge.org/9780521849890

First published 2009

A catalogue record for this publication is available from the British Library

Library of Congress Cataloging-in-Publication data
Gonnet, G. H. (Gaston H.)
Scientific computation / Gaston Gonnet, Ralf Scholl.
p.   cm.
Includes index.
ISBN 978-0-521-84989-0 (alk. paper)
1. Computer science – Mathematics.   2. Science – Data processing.
3. Data structures (Computer science)   4. Bioinformatics.   I. Scholl, Ralf.
II. Title.
QA76.9.M35G64   2009
501′.51 – dc22   2009011758

ISBN   978-0-521-84989-0   Hardback

Additional resources for this publication at www.cambridge.org/9780521849890

Contents

Preface

This book contains the material used in our advanced course on Scientific Computation at the ETH (Eidgenössische Technische Hochschule), Zürich, Switzerland. This course is a third year course in the Department of Computer Science (Informatik).

The material is presented in a non-traditional way by solving several, reasonably realistic problems. Optimization techniques, modelling, simulation and more mathematical subjects are introduced as needed to solve these problems. We believe that this helps students understand the methods in some of their appropriate contexts, and hence gives a more comprehensive view of scientific computation.

When necessary we have indicated in the outside margin the kind of material.

BASIC   The very basics necessary to understand the following material.

ADVANCED   Advanced material going beyond the average scope of the book.

EXAMPLE   Worked out examples.

PRACTICAL NOTE   Practical notes, usually derived from our experience with scientific computation.

BACKGROUND   Background material which explains the history or rationale of a given topic or research area.

Synopsis

This book consists of eight chapters and two appendixes. In each of the eight chapters one problem is treated. Appendix A contains mathematical background on minimization methods, and Appendix B lists online resources.

Here is an overview of the problems and the academic goals addressed in the various chapters.

Chapter	Problem	Academic goals	Algorithms covered
1	position of an aircraft	non-linear least squares (LS), errors, confidence analysis	analytic solution
2	car and computer replacement	modelling, LS for exponentials, piecewise approximation	analytic solution
3	secondary structure prediction of proteins	modelling LS with SVD	singular value decomposition, eigenvalue decomposition
4	secondary structure prediction of proteins	modelling LS with best basis, confidence intervals	finding discrete local minima, early abort (EA), discrete steepest descent
5	SSP of proteins	nearest neighbors (NN)	NN trees, k-d trees, clustering
6	SSP of proteins	linear programming	simplex algorithm
7	stock market prediction	modelling, dynamic programming, parameter optimization, discrete function minimization, confidence analysis	dynamic programming, PiecewiseConstMax, random directions, spiral search, shrinking hypercube
8	phylogenetic tree construction, string matching	Markovian evolution, dynamic programming, discrete optimization, evolutionary algorithms	PAM distance estimation, global/local alignment, UPGMA, WPGMA, 4-optim, 5-optim, general parsimony, perfect phylogeny, CHE methods
Appendix A	mathematical background	mathematical methods of function minimization	golden section search, Brent's method, parabolic interpolation, steepest descent, random directions, Newton's method, spectral method

Abbreviations

AA	amino acid
BB	best basis
CHE	constructor–heuristic–evaluator
DME	distance measuring equipment
EA	early abort
ERP	equipment replacement problem
EV(D)	eigenvalue (decomposition)
EVA	evolutionary algorithm
k-d	k-dimensional
LP	linear programming
LS	(method of) least squares
MSA	multiple sequence alignment
NN	nearest neighbors
OTS	optimal transaction sequence
PAM	point accepted mutations (obsolete, only the abbreviation PAM is in use)
RD	random directions
SD	steepest descent
SM	spectral method
SVD	singular value decomposition
UPGMA	unweighted pair group method with arithmetic mean
VOR	very high frequency omnirange
WPGMA	weighted pair group method with arithmetic mean

1 Determination of the accurate location of an aircraft

In this chapter we want to find the most accurate location of an aircraft using information from beacons. This is essentially the same problem that a GPS system or a cellular phone has to solve.

Topics
- Non-linear least squares
- Statistical errors
- Function minimization
- Sensitivity analysis

The chapter is organized as follows:

- in Section 1.1 the problem is described,
- in Section 1.2 we will show how to model this problem mathematically,
- in Section 1.3 we will solve it analytically, and
- in Section 1.4 we will explore methods to analyze the solution we have found.

In Appendix A we explore different minimization methods.

1.1 Introduction

Figure 1.1 illustrates a simplified typical situation of navigation with modern aircraft. The airplane is in an unknown position and receives signals from various beacons. Every signal from the beacons is assumed to contain some error. The main purpose of this problem is to develop a method for computing the most likely position of the aircraft based on all the information available.

We distinguish two kinds of beacons: very high frequency omnirange (VOR) and distance measuring equipment (DME). The VOR beacons allow the airplane to read

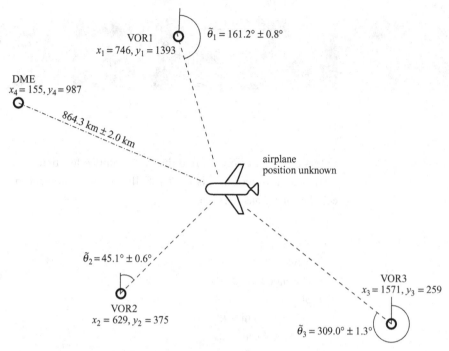

VOR1
$x_1 = 746, y_1 = 1393$

$\tilde{\theta}_1 = 161.2° \pm 0.8°$

DME
$x_4 = 155, y_4 = 987$

864.3 km ± 2.0 km

airplane
position unknown

$\tilde{\theta}_2 = 45.1° \pm 0.6°$

VOR3
$x_3 = 1571, y_3 = 259$

VOR2
$x_2 = 629, y_2 = 375$

$\tilde{\theta}_3 = 309.0° \pm 1.3°$

Figure 1.1 Example of an aircraft and four beacons.

the angle from which the signal is coming. In other words, θ_1, θ_2 and θ_3 are known to the airplane. The DME beacon, using a signal that is sent and bounced back, allows the distance from the airplane to the beacon to be measured. In this example the distance is 864.3 km $\pm$ 2.0 km.

Each of the measurements is given with an estimate of its error. The standard notation for measurements and errors is $m \pm n$. This means that the true value being measured lies between $m - n$ and $m + n$. Different disciplines have different interpretations for the statement "lies between." It may mean an absolute statement, i.e. the true value is always between the two bounds, or a statistical statement, i.e. the true value lies within the two bounds $z\%$ of the time. It is also common to assume that the error has a normal distribution, with average m and standard deviation n. For our analysis, it does not matter which definition of the error range is used, provided that all the measures use the same one.

We will simplify the problem by considering it in two dimensions only. That is, we will not consider the altitude, which could be read from other instruments and would unnecessarily complicate this example. We will denote by x and y the unknown coordinates of the aircraft.

The input data are summarized in the following table.

	x coordinate	y coordinate	value	error
VOR1	$x_1 = 746$	$y_1 = 1393$	$\tilde{\theta}_1 = 161.2°$	$\tilde{\sigma}_1 = 0.8°$
VOR2	$x_2 = 629$	$y_2 = 375$	$\tilde{\theta}_2 = 45.1°$	$\tilde{\sigma}_2 = 0.6°$
VOR3	$x_3 = 1571$	$y_3 = 259$	$\tilde{\theta}_3 = 309.0°$	$\tilde{\sigma}_3 = 1.3°$
DME	$x_4 = 155$	$y_4 = 987$	$\tilde{d}_4 = 864.3$ km	$\tilde{\sigma}_4 = 2.0$ km
aircraft	x	y		

It is easy to see, that unless we are in a pathological situation, any pair of two VOR/DME readouts will give enough information to compute x and y. If the measurements were exact, the problem would be overdetermined with more than two readouts. Since the measures are not exact, we want to compute x and y using all the information available and, hopefully, obtain a more accurate answer.

The standard measure of angles in aviation is clockwise from North in degrees. This is different from trigonometry, which uses counterclockwise from East in radians. Hence care has to be taken with the conversion from degrees to radians.

We do this step first and get the following results.

	x coordinate	y coordinate	value	error
VOR1	$x_1 = 746$	$y_1 = 1393$	$\tilde{\theta}_1 = 5.0405$ rad	$\sigma_1 = 0.014$ rad
VOR2	$x_2 = 629$	$y_2 = 375$	$\tilde{\theta}_2 = 0.784$ rad	$\sigma_2 = 0.0105$ rad
VOR3	$x_3 = 1571$	$y_3 = 259$	$\tilde{\theta}_3 = 2.461$ rad	$\sigma_3 = 0.023$ rad
DME	$x_4 = 155$	$y_4 = 987$	$\tilde{d}_4 = 864.3$ km	$\sigma_4 = 2.0$ km
aircraft	x	y		

Definition of best approximation Find the aircraft position (x, y) which minimizes the error in the following way.

- Regard the total error ε as a vector of all the measurement errors. This vector of errors contains the error of each measurement for a value of (x, y); in our example it has four components.
- As norm of this vector we use the usual euclidean norm $\|\varepsilon\|_2$ which is defined in our example as $\|\varepsilon\|_2 := \sqrt{\sum_{i=1}^{4} \varepsilon_i^2}$. (We could use other norms instead, for example $\|\varepsilon\|_{\max} := \max(|\varepsilon_i|)$.)
- Find the (x, y) which minimizes the norm of the total error, hence for $\|\varepsilon\|_2$ use the method of least squares (LS).

1.2 Modelling the problem as a least squares problem

Under the assumption that the errors are normally distributed, it is completely appropriate to solve the problem of locating x and y by minimizing the sum of the squares of the errors. On the other hand, if we do not know anything about the distribution of the individual errors, minimizing the sum of their squares has a simple geometrical interpretation, the euclidean norm, which is often a good idea. So, without further discussion, we will pose the problem as a least squares (LS) problem.

We can relate the unknown exact position (x, y) of the airplane with the given VOR positions (x_i, y_i) for $i = 1, \ldots, 3$ by:

$$\tan(\theta_i) = \frac{x - x_i}{y - y_i} \tag{1.1}$$

where θ_i is the angle to the unknown exact airplane position. Using the DME position (x_4, y_4), we get the equation

$$d_4 = \sqrt{(x - x_4)^2 + (y - y_4)^2} \tag{1.2}$$

for d_4, the distance to the unknown exact position of the airplane.

Each of the measures is subject to an error. We will call these errors for the different measures ε_i. Hence $\varepsilon_i = \theta_i - \tilde{\theta}_i$, where θ_i is the real value, $\tilde{\theta}_i$ is the actual measure of the value and ε_i is the measurement error. (If the measure is given as $m \pm n$, $\tilde{\theta}_i = m$.) For example, $\theta_1 = 161.2 + \varepsilon_1$, and we mean θ_1 is the exact angle and ε_1 is the error of the actual measurement.

Now the above Equations (1.1) and (1.2) can be written as:

$$\tan(\theta_i) = \tan(\tilde{\theta}_i + \varepsilon_i) = \frac{x - x_i}{y - y_i} \qquad \text{for } i = 1, \ldots, 3 \tag{1.3}$$

$$d_4 = \tilde{d}_4 + \varepsilon_4 = \sqrt{(x - x_4)^2 + (y - y_4)^2}. \tag{1.4}$$

All the variables are related by this system of four equations in the six unknowns x, y, ε_1, ε_2, ε_3, ε_4. It is normally underdetermined, so we cannot determine the exact position. Instead we are going to determine the solution which minimizes the norm of the total error $\varepsilon = (\varepsilon_1, \varepsilon_2, \varepsilon_3, \varepsilon_4)$.

The errors ε_i, when viewed as random errors, have a known average, namely 0 (since the measurement instruments are typically unbiased), and a known variance or standard deviation. For example, ε_1 has average 0 and a standard deviation 0.014 rad. In general for a measure $m_i \pm \sigma_i$ the associated error has average 0 and standard deviation σ_i or variance σ_i^2. If the errors are assumed to be normally distributed, then they have a normal (gaussian) distribution $\mathcal{N}(0, \sigma_i^2)$.

When we minimize the norm of ε it should be done on similarly distributed variables ε_i. (We want to compare apples to apples and not apples to oranges.) To achieve this, we will divide each ε_i by its standard deviation σ_i. The normalized errors ε_i/σ_i have distribution $\mathcal{N}(0, 1)$.

So, since we are using the euclidean norm, that is, we want to minimize the length of the "normalized" error-vector $(\varepsilon_1/\sigma_1, \varepsilon_2/\sigma_2, \varepsilon_3/\sigma_3, \varepsilon_4/\sigma_4)$, we have to minimize $\sum (\varepsilon_i/\sigma_i)^2$. Typically every error ε_i appears in only one equation, and hence it is easy to solve for it, e.g.

$$\varepsilon_4 = \sqrt{(x - x_4)^2 + (y - y_4)^2} - \tilde{d}_4 = \sqrt{(x - 155)^2 + (y - 987)^2} - 864.3.$$

Inverting the equations for $\varepsilon_{1,\ldots,3}$ which contain the tangent function poses a small technical problem because of the periodicity of the tangent function. $\tan(\hat{\theta}_i + \varepsilon_i) = (x - x_i)/(y - y_i)$ for $i = 1, \ldots, 3$ is always correct, but inverting, we obtain $\varepsilon_i = \arctan((x - x_i)/(y - y_i)) - \hat{\theta}_i + k\pi$ with $k \in \mathbb{Z}$. Inverting trigonometric equations with a computer algebra system like Maple will normally return the principal value, that is a value between $-\pi/2$ and $\pi/2$, which may be in the wrong quadrant.

PRACTICAL NOTE This brings two problems, one of them trivial, the second one more subtle. The trivial problem is how to convert aviation angles, which after normalization will be in the range from zero to 2π, to the range $-\pi$ to π. This is done by subtracting 2π from the angle if it exceeds π.

The second problem is that arctan returns values between $-\pi/2$ and $\pi/2$. This means that opposite directions, for example the angles $135°$ and $315°$, are indistinguishable. This may result in an equation that cannot be satisfied, or if the angles are reduced to be in the arctan range, then multiple, spurious solutions are possible. To correct this problem we analyze the signs of $x - x_i$ and $y - y_i$ to determine the correct direction, which is called quadrant analysis. This is a well known and common problem, and the function `arctan` with two arguments in Maple (`atan2` in C and Java) does the quadrant analysis and returns a value between $-\pi$ and π, resolving the problem of opposite directions (`arctan` with two arguments also resolves the problem of $y - y_i = 0$ which should return $\pi/2$ or $-\pi/2$ but could cause a division by zero).

Equation (1.3) should be rewritten as:

$$\varepsilon_i = \texttt{arctan}(x - x_i, y - y_i) - \tilde{\theta}_i.$$

Finally the sum of squares that we want to minimize in our example is

$$S(x, y) = \sum_{i=1}^{3} \left(\frac{\texttt{arctan}(x - x_i, y - y_i) - \tilde{\theta}_i}{\sigma_i} \right)^2 + \left(\frac{\sqrt{(x - x_4)^2 + (y - y_4)^2} - \tilde{d}_4}{\sigma_4} \right)^2.$$

This problem is non-linear in its unknowns x and y as x and y are simultaneously arguments of an arctan function and inside a square root function. This means that an explicit solution of the least squares problem is unlikely to exist and we will have to use numerical solutions.

See the interactive exercise "Least squares."

1.2.1 How to solve a mixed problem: equation system and minimization

BASIC

Suppose we want to minimize a function subject to constraints. This can be viewed as solving a minimization jointly with a set (or system) of equations, the constraints. Let $eq_1, eq_2, \ldots, eq_i$ be the equations (constraints), and let $f(x_1, x_2, \ldots, x_k)$ be the function to be minimized.

A simple procedure to solve this problem is by substitution. It consists of the following steps.

(i) Choose an equation eq_j and an unknown in eq_j. (This choice should be made for the equation/unknown which is easiest to solve. The easiest case is solving for an unknown which does not appear in any other equation, in our example the ε_j.)

(ii) Solve eq_j for the unknown and substitute the value for this unknown in f and in all other remaining equations.

(iii) Remove eq_j from the set of equations.

When no more constraints are left we can minimize f in terms of the unknowns that remain. The substituted variables can be computed from the minimal solution by backsubstitution, if desired.

Notice that in our case, normally each error is associated with a measure and each measure gives a constraint (equation). This error will not appear in any other measure/equation. So we have a way of solving all equations easily by substitution.

See the interactive exercise "Measurement."

1.3 Solving the non-linear least squares problem

To solve this problem we will use the computer algebra system Maple, since we need to do some symbolic as well as numerical computations. First we define the input data. We use the vectors X and Y to store the beacon coordinates and x and y for the unknown coordinates of the airplane.

```
>   theta := array([161.2,45.10,309.0]);
>   sigma := array([0.8,0.6,1.3,2.0]);
>   X := array([746,629,1571,155]);
>   Y := array([1393,375,259,987]);
>   d4 := 864.3;
```

$$\theta := [161.2,\ 45.10,\ 309.0]$$

$$\sigma := [0.8,\ 0.6,\ 1.3,\ 2.0]$$

$$X := [746,\ 629,\ 1571,\ 155]$$

$$Y := [1393,\ 375,\ 259,\ 987]$$

$$d_4 := 864.3$$

The angles and the standard deviation of angles have to be converted to radians, as described earlier. These are the calculations which were performed:

```
>   for j from 1 to 3 do
>       theta[j] := evalf(2*Pi*theta[j]/360);
>       if theta[j] > evalf(Pi) then
>           theta[j] := theta[j] - evalf(2*Pi)
>       fi;
>       sigma[j] := evalf(2*Pi*sigma[j] / 360);
>   od:
>   print(theta);
>   print(sigma);
```

$$[2.813470755,\ 0.7871434929,\ -0.890117918]$$

$$[0.01396263402,\ 0.01047197551,\ 0.02268928028,\ 2.0]$$

We are now ready to construct the sum of squares.

```
>   S := sum(((arctan(x-X[i],y-Y[i])-theta[i])/sigma[i])^2, i=1..3)+
>              +((((x-X[4])^2+(y-Y[4])^2)^(1/2)-d4)/sigma[4])^2;
```

$$S := 5129.384919\,(\arctan(x-746,\ y-1393) - 2.813470755)^2$$
$$+\, 9118.906531\,(\arctan(x-629,\ y-375) - 0.7871434929)^2$$
$$+\, 1942.488964\,(\arctan(x-1571,\ y-259) + 0.890117918)^2$$
$$+\, 0.2500000000\,(\sqrt{(x-155)^2 + (y-987)^2} - 864.3)^2$$

Next we solve numerically for the derivatives equated to zero. In Maple, fsolve is a basic system function which solves an equation or system of equations numerically.[1]

```
>   sol := fsolve({diff(S,x)=0, diff(S,y)=0},{x=750,y=950});
```

$$sol := \{x = 978.3070298,\ y = 723.9837773\}$$

---

[1] $(x = 750, y = 950)$ is an initial guess to the solver fsolve. It will try to find a solution starting from this point. This is useful for two reasons, first it improves the efficiency of the solver and secondly it increases the chances that we converge to a minimum (rather than other places where the derivative is zero like maxima or saddle points).

A solution has been found and it is definitely in the region that we expect it to be. The first measure of success or failure of the approximation is to examine the residues of the least squares approximation. Under the assumption that the errors are normally distributed, this will be the sum of the squares of the four $\mathcal{N}(0, 1)$ variables. Note that

$$E[x_1^2 + \cdots + x_4^2] = 4 \quad \text{if} \quad x_i \sim \mathcal{N}(0, 1)$$

but for the least squares sum $S = \sum_i \varepsilon_i^2$ we choose *optimal x, y* when we minimize S. Therefore we expect

$$E[S_{\min}] = 2$$

since this system has only two degrees of freedom.[2] The norm squared of the error is:

```
>   S0 := evalf(subs(sol, S));
```

$$S_0 := 0.6684712637$$

This value is smaller than 2, and hence it indicates that either we are lucky, or the estimates for the errors were too pessimistic. In either case, this is good news for the quality of the approximation. This together with the eigenvalue analysis from the next section guarantees that we have found the right solution.

1.4 Error analysis/confidence analysis

A plain numerical answer, like the result from `fsolve` above, is not enough. We would like to know more about our result, in particular its confidence. Confidence analysis establishes the relation between a region of values T and the probability that the correct answer lies in this region. In our case we are looking for a range of values which are a "reasonable" answer for our problem, see Figure 1.2:

$$P[\text{given } T, \text{ the real answer is outside } T] = \begin{cases} 10^{-6} & \text{very precise} \\ 0.001 & \text{precise} \\ 0.01 & \\ 0.05 & \text{reasonable} \\ p & \text{in general.} \end{cases}$$

---

[2] The discussion on the degrees of freedom goes beyond the scope of this book. A simple rule of thumb is that if our minimization, after substitution for all the constraints, has k variables left and we have n errors, then the degrees of freedom are $n - k$. In this case: $4 - 2 = 2$.

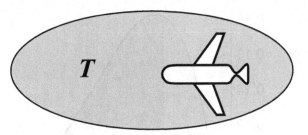

Figure 1.2   **The exact position of the airplane is not known, we can only compute regions T where the airplane is likely to be.**

If the errors are assumed to be normally distributed, the sum of the squares of the errors has a known distribution, called the χ^2-distribution (read: chi-square distribution).

We will work under the assumption that the errors of the measures have a normal distribution. The distribution of $S(x, y)$ is a χ^2-distribution with four degrees of freedom, since it is a sum of squares of four variables which are $\mathcal{N}(0, 1)$ distributed.[3]

For a given confidence level we can bound the value of χ^2 and hence bound the solution (x, y), for example by $S(x, y) \leq v$ with $\Pr(\{\chi_4^2 \leq v\}) =$ confidence. This inequality defines an x, y area T, which is an area where the airplane will be located with the given confidence. (Note: the bigger the confidence, i.e. the lesser the probability of an error, the bigger is the area.)

Knowing its distribution allows us to define a confidence interval for the airplane position. Suppose that we are interested in a 95% confidence interval, then $S(x, y) < v \approx 9.4877$ (Figure 1.3), where this value is obtained from the inverse of the cumulative (icdf) of the χ^2-distribution. In Maple this is computed by:

```
>   stats[statevalf,icdf,chisquare[4]](0.95);
```

$$9.487729037$$

The inequality $S(x, y) < 9.4877$ defines an area which contains the true values of x and y with probability 95%. We can draw three areas for three different confidence intervals, e.g. 50%, 95%, 99.9%, all of which are reference values in statistical computations. Notice that the larger the confidence, the larger the ellipse (see Figure 1.4). And see the interactive exercise "Quality control."

---

[3] We do not regard S as a function of (x, y) in this case, since (x, y) is kept fixed (it should be the true position of the airplane). Instead we regard S as a random variable, since it still depends on the *four* measurements, which are random variables. Since these measurements contain independent, normally distributed errors $\varepsilon_1, \ldots, \varepsilon_4$, S has a χ^2-distribution with four degrees of freedom.

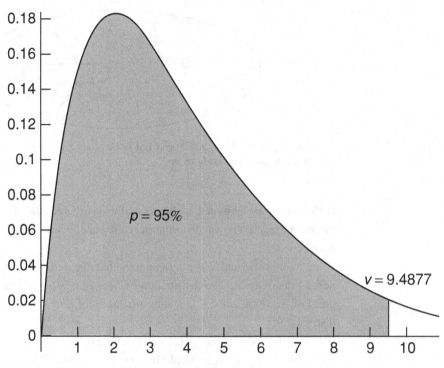

Figure 1.3 The probability density function of a χ^2-distribution with four degrees of freedom.

1.4.1 An approximation for a complicated S

For complicated $S(x, y)$, finding the area may be computationally difficult. So we will show how to use an approximation. We will expand the sum of the squares of the errors as a Taylor series around the minimum and neglect terms of third and higher order. So let $S(x, y) = S(p)$ be the sum of squares, which we will define as a function of the position vector $p = (x, y)^T$. Let p_0 be the solution of the least squares problem. Then the three-term Taylor series around p_0 is

$$S(p) = S(p_0) + S'(p_0)(p - p_0) + \frac{1}{2}(p - p_0)^T S''(p_0)(p - p_0) + \mathcal{O}(\|p - p_0\|^3).$$

The gradient of S, $S'(p) = (S_x, S_y)^T$, is always zero at the minimum, and a numerical check shows that the gradient in our example is indeed within rounding error of $(0, 0)$:

```
>   S1 := evalf(subs(sol,linalg[grad](S,[x,y])));
```
$$S_1 := [0.36 \times 10^{-7}, -0.6 \times 10^{-8}]$$

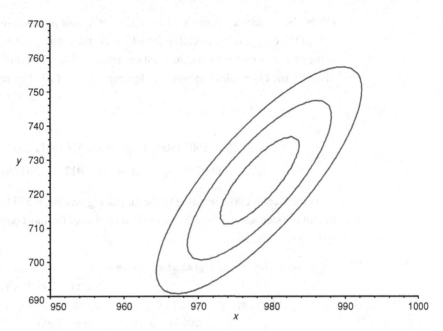

Figure 1.4 **Ellipses of the expected locations for probabilities 50% (innermost), 95% (middle) and 99.9% (outermost curve).**

So the expression for S ignoring higher order terms simplifies to

$$S(\boldsymbol{p}) = S(\boldsymbol{p}_0) + \frac{1}{2}(\boldsymbol{p} - \boldsymbol{p}_0)^{\mathrm{T}} S''(\boldsymbol{p}_0)(\boldsymbol{p} - \boldsymbol{p}_0).$$

In Maple the computation of the approximation to S in terms of the unknowns x and y is carried out by first computing the Hessian (the matrix with the second derivatives of S), evaluating it at the minimum and then computing the quadratic form .

```
>   S2 := evalf(subs(sol,linalg[hessian](S,[x,y])));
```

$$S_2 := \begin{bmatrix} 0.5118424681 & -0.1726069278 \\ -0.1726069278 & 0.09026159227 \end{bmatrix}$$

```
>   pmp0 := [x-subs(sol,x),y-subs(sol,y)];
```

$$\mathrm{pmp}_0 := [x - 978.3070298, \; y - 723.9837773]$$

```
>   Sapprox := S0 + (1/2)*evalm(transpose(pmp0) &* S2 &* pmp0);
```

$$S_{\mathrm{approx}} := 0.6684712637$$

$$+ \frac{1}{2}(0.5118424681\,x - 375.7744691 - 0.1726069278\,y)(x - 978.3070298)$$

$$+ \frac{1}{2}(-0.1726069278\,x + 103.5146424 + 0.09026159227\,y)(y - 723.9837773)$$

(Note: the symbol &* denotes vector and matrix multiplication in Maple.) The major axis of the ellipse is the direction for which the uncertainty is largest. The minor axis is the one for which we have the least uncertainty. The exact direction of both axes is given by the eigenvalue/eigenvector decomposition of S'' (Figure 1.4).

```
>   ev := linalg[eigenvects](S2);
```

$$ev := [0.0286079804, 1, \{[-0.3363764637, -0.9417276012]\}],$$
$$[0.5734960799, 1, \{[-0.9417276012, 0.3363764637]\}]$$

The eigenvector corresponding to the largest eigenvalue, $0.5734\ldots$, gives the direction of the steepest climb of $S(p)$, or the direction of the most confidence, in this case $(-0.941\ldots, 0.336\ldots)$.

```
>   ellips := {seq(stats[statevalf, icdf, chisquare[4]](c) =
>                             Sapprox, c=[0.5,0.95,0.999])}:
>   plots[implicitplot](ellips, x=950..1000, y=690..770,
>                 grid=[50,50],  view=[950..1000, 690..770]);
```

The eigenvector corresponding to the smallest eigenvalue, $0.0286\ldots$, gives the direction of the least confidence, which is obviously perpendicular to the previous direction. For this particular example, we see that the DME and VOR2 probably contribute the most information (the DME beacon had the smallest relative error), and hence the shape and orientation of the ellipses.

(PRACTICAL NOTE) **Note** The larger the confidence (i.e. the less probable an error), the larger the argument of the χ^2-function has to be, and the larger the ellipsoid will be.

Computing the eigenvalues has one additional advantage. When we solve for the derivatives equal to zero, we could find a minimum, or a maximum or a saddle point. The proper way of checking that we obtained a minimum is to verify that the Hessian matrix is positive definite, i.e. all eigenvalues are positive. Inspecting the eigenvalues gives even more information: if all the eigenvalues are positive, then we have a minimum, if all are negative then we have a maximum, and if the eigenvalues have mixed signs, we are at a saddle point. Since both eigenvalues are positive, we have confirmed that we have found a minimum. The procedure described in this problem can be extended trivially to any number of dimensions (i.e. unknown parameters), see Appendix A, Section A1.3.

1.4.2 Interactive exercise

For a situation as in Figure 1.1 you can change the positions of the beacons, their accuracy and their measurement values. As a result you will get the estimated airplane position and a plot of the resulting confidence region. A cross (+) in the resulting plot marks the approximated best airplane position obtained by the least squares method. The confidence region corresponding to the first specified confidence level is depicted as a red ellipse, the second specified level is shown as a blue ellipse, the third as a cyan (bluegreen) ellipse and the fourth as magenta (purple). If more confidence levels are specified these colors are repeated. Be aware that some confidence regions can be empty, depending on the quality of the estimation. See exercise 3 in the online version!

FURTHER READING

Sections 1.1–1.4 are based on

W. Gander and J. Hřebíček, *Solving Problems in Scientific Computing using Maple and Matlab*, Springer Verlag, 2004, Chapter 26.

When to replace equipment

Topics
- Basics of modelling and simulation
- Function approximation

The following is a practical problem which we subject to mathematical analysis and optimization, that of replacing equipment (ERP), for example replacing your car or your computer. We will analyze these problems in detail, and find the simplest closed form solutions.

On the other hand, these types of problems are easily solved by simulation, and this is the tool which should be used when the problems have enough details to make them more realistic.

So we have a choice between closed form solutions and simulation: simulations have greater accuracy in the sense that they reflect reality more closely, closed form solutions indicate functional dependence on the parameters, giving structural insights. Both are very important tools.

2.1 The replacement paradox

The replacement paradox refers to the familiar situation that one faces when equipment has aged and it may be time to replace it. At a given time two choices are available: continue using the old equipment, which has lower performance and higher maintenance cost, or buy new equipment which will have better performance and lower maintenance, but a higher initial capital expenditure. The paradox arises from the fact that if we look only a short period of time into the future the decision is always bent towards keeping the old equipment, never replacing it. The maintenance and repair costs for the immediate future are far less than the replacement cost. Globally however there is an optimal time for replacement, but locally it is difficult (or impossible) to justify the replacement.

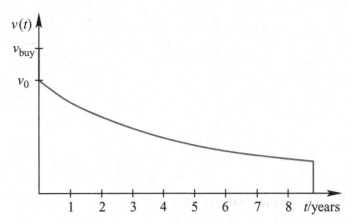

Figure 2.1
The value function $v(t)$.

2.2 Car replacement

We would like to describe the main cost functions associated with a car (which are also applicable to other equipment).

First of all we have the value function $v(t)$, the value of the car at time t. This is what we have to pay for the car and what we would get for it if we sold it. Typically this curve is an exponential decay with a jump at the beginning and an abrupt end (Figure 2.1).

The initial jump downwards represents the depreciation caused by various things like "driving off the lot," taxes, sales commissions etc. In other words, any cost that we do not recover if we buy and sell immediately. The final drop is due to the fact that at some point the car becomes impossible to sell (because of regulations, market conditions etc.) and its value drops to zero. The remainder of the curve is usually assumed to be an exponential decay which can be approximated from real data or estimated from statements like "cars depreciate 15% of their value per year."

If the above statement is taken verbatim, then the value of the car is

$$v(t) = \begin{cases} v_{buy} & \text{for } t = 0 \\ v_0 \cdot 0.85^t & \text{for } 0 < t < t_{max} \\ 0 & \text{for } t \geq t_{max} \end{cases}$$

when t is measured in years.

Alternatively, to estimate this function from data, we can take two values v_a at $t = t_a$ and v_b at $t = t_b$ (the more years in between the better).

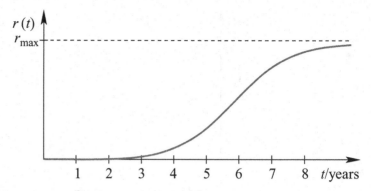

Figure 2.2   The repair cost function.

Let us say $v_a > v_b$, hence $t_b > t_a$. Then

$$v(t) = v_a \left(\frac{v_b}{v_a} \right)^{\frac{t - t_a}{t_b - t_a}}$$

is an exponential decay, which satisfies the given conditions. If we use the maximum period, that is from $t = 0$ up to t_{max} (t_{max} being the time of the drop in value to 0) then

$$v(t) = v_0 \left(\frac{v_{max}}{v_0} \right)^{\frac{t}{t_{max}}}.$$

With more data points we can approximate the exponential by least squares, see Section 2.2.3.

The next important function is the cost of repairs $r(t)$. This is an average of the cost per time (or the expected cost per time) to keep the car (equipment) in working order. This function is very low or zero at the beginning since the equipment is new and not likely to break, and/or is under guarantee, then the function grows and eventually reaches an asymptotic maximum value (Figure 2.2).

The cost of repairs should also include costs associated with failure of the equipment such as alternative transportation, etc. This function is not easy to model mathematically. Often we will have to approximate it with a few straight lines using a "piecewise linear approximation."

The problem of considering inflation may be taken care of by working with adjusted currency, or in the case of equipment with a shorter life, by just ignoring inflation. Adjusting currency to today's values is quite simple, we just multiply the actual values by the inflation accrued since that time. For example, assume an inflation rate of 3%, so $1000 spent in 1998 have a value equivalent to $1000 \cdot 1.03^{10} = \$1343.91$ at 2008 values. Hence we are not going to consider inflation any longer.

To these two basic functions, value $v(t)$ and repair costs $r(t)$, we could add other cost functions. The cost of operation per unit of time $o(t)$ is very relevant (gasoline,

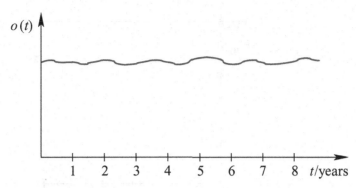

Figure 2.3   The cost of operations function.

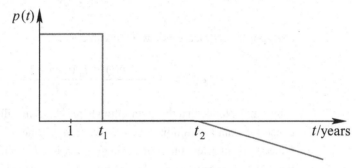

Figure 2.4   The pleasure function.

maintenance, insurance, road taxes, etc.) and it may change from model to model. The composite of all these costs is rather flat (Figure 2.3), as gas costs go up, insurance goes down. Other values depend on how much you drive, which should remain about constant over the years in our simple model; if not, the model has to be adapted accordingly. For example, the rising gas prices over recent years could be modelled with a rising function $o(t)$, leading to earlier replacement times, provided newer car models consume less gas. We won't model this.

If this curve $o(t)$ is flat (constant), it will not affect the decision to replace the car. It is easy to see that any constant cost cannot affect the replacement time.

Another interesting factor to consider is the pleasure $p(t)$ of driving a new/old car. We could assign a value to this by estimating how much we are willing to pay for this pleasure. It will become negative for an aggravation. Typically it will look like the graph shown in Figure 2.4.

At some point t_1 (say, two years) the excitement of a new car wears out. Later at t_2 (say, five years) the aggravation costs kick in.

At this point we can state the optimization problem that we want to solve. We want to find an optimal t so that all the costs for the period t divided by t, i.e. the average

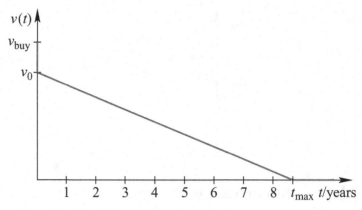

Figure 2.5 The value function.

costs per time $a(t)$, are as low as possible, i.e.

$$a(t) := \frac{v_{\text{buy}} - v(t) + \int_0^t (r(x) + o(x) - p(x)) \, dx}{t}$$

is minimal. Notice that the assumption here is that we will replace the equipment, and hence we want to minimize the cost per unit of time. If we are not going to replace the equipment after some time, the problem is different: we then want to minimize the cost up to a final time t_f. In that case replacements close to t_f are much less likely to occur, which justifies in economic terms why older people drive older cars.

2.2.1 Simulation

Optimizing the replacement time by simulation means computing all possible values of $a(t)$ and selecting the one which is best. At this point we could use numerical minimization but in this particular application where the granularity is either one year or (at best) one month, direct computation of all possibilities is the easiest solution.

This is the simplest form of simulation, and because we have a single unknown (the replacement time) it is very efficient. See Section 2.3.1 for more details.

2.2.2 Linear approximations

The simplest model of value $v(t)$ and repair cost $r(t)$ is to use linear functions, i.e. straight lines. This is of course a poor approximation but it is so simple that it allows a closed form solution. The value $v(t)$ becomes a straight line (Figure 2.5) between

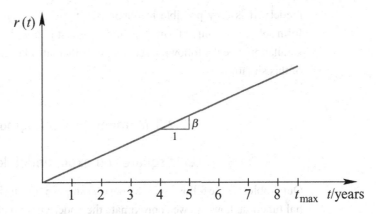

Figure 2.6 The repair cost function.

v_0 the reselling price at $t = 0$ and zero at t_{max}, i.e.

$$v(t) = \begin{cases} v_{buy} & \text{for} \quad t = 0 \\ \left(1 - \frac{t}{t_{max}}\right) v_0 & \text{for} \quad 0 < t \le t_{max} \\ 0 & \text{for} \quad t > t_{max}. \end{cases}$$

The cost of repairs will also be a straight line (Figure 2.6) going from 0 onwards:

$$r(t) = \beta t$$

where β is the increment in repair cost per year. All the other cost functions will be ignored, since we assume them to be practically constant. Under these simplifications the total costs per time-unit are given by

$$a(t) = \frac{v_{buy} - \left(1 - \frac{t}{t_{max}}\right) v_0 + \int_0^t \beta x \, dx}{t}$$

$$= \frac{v_{buy} - v_0}{t} + \frac{v_0}{t_{max}} + \frac{\beta t}{2}.$$

This is simple enough that we can find its minimum by equating the derivatives with respect to zero:

$$a'(t) = -\frac{v_{buy} - v_0}{t^2} + \frac{\beta}{2} = 0 \qquad \Rightarrow$$

$$t = \sqrt{\frac{2 \left(v_{buy} - v_0\right)}{\beta}} \qquad \text{if } t \le t_{max}.$$

Notice that the formula does not depend on t_{max}, hence t_{max} is a parameter that we do not need to know to be able to compute the optimal time for replacement. However, the answer has to be $t \le t_{max}$ for the equations to be valid. Only the initial jump and the cost of repairs influence the optimal time of replacement (in this very simple model). It is only possible to obtain this type of information/insight from closed form solutions (simple formulas), and is almost impossible to obtain from numerical simulation. See the following section on mathematical background for the full details of this situation.

2.2.3 Mathematical background

Least squares on exponential decays

Our problem is the following: if we have n sample points $v_i, i = 1 \ldots n$, of an exponential function, how can we approximate the underlying function best? Mathematically this is expressed as

$$v_i \approx a_0 \alpha^{t_i} \quad \text{for} \quad i = 1 \ldots n$$

where a_0 and α are the parameters to be determined.

If we want this approximation to be by least squares (LS) then we should minimize

$$\sum_{i=1}^{n} \left(v_i - a_0 \alpha^{t_i} \right)^2 .$$

Calculating the derivative with respect to a_0 we get

$$2 \sum_{i=1}^{n} \left(-\alpha^{t_i} \cdot \left(v_i - a_0 \alpha^{t_i} \right) \right) = 0$$

or

$$a_0 = \frac{\displaystyle\sum_{i=1}^{n} \alpha^{t_i} v_i}{\displaystyle\sum_{i=1}^{n} \alpha^{2t_i}} . \tag{2.1}$$

However, the derivative with respect to α will give a non-linear equation to solve:

$$2 \sum_{i=1}^{n} \left(-a_0 t_i \alpha^{t_i - 1} \cdot \left(v_i - a_0 \alpha^{t_i} \right) \right) = 0$$

which simplifies to

$$\sum_{i=1}^{n} t_i \alpha^{t_i - 1} \cdot \left(v_i - a_0 \alpha^{t_i} \right) = 0$$

or using Equation (2.1)

$$\sum_{i=1}^{n} v_i t_i \alpha^{t_i-1} = \frac{\displaystyle\sum_{i=1}^{n} v_i \alpha^{t_i} \sum_{i=1}^{n} t_i \alpha^{2t_i-1}}{\displaystyle\sum_{i=1}^{n} \alpha^{2t_i}}.$$

This type of equation can only be solved numerically, there is no closed form for the solution.

Sometimes this problem is solved in a different way. Instead of solving $v_i \approx a_0 \alpha^{t_i}$ we approximate the logarithms of the values by least squares (which is *not* the same problem):

$$\ln(v_i) \approx \ln(a_0 \alpha^{t_i}) = \ln(a_0) + t_i \ln(\alpha).$$

In this case the problem becomes the minimization of

$$\sum_{i=1}^{n} (\ln v_i - \ln a_0 - t_i \ln(\alpha))^2,$$

which is now a *linear* least squares problem. The problem is linear in $\ln(a_0)$ and $\ln(\alpha)$, which is enough to solve it exactly. To compute a_0 and α we then compute the exponentials. Let us do this now!

By computing the derivatives we get a system of two linear equations in $\ln(a_0)$ and $\ln(\alpha)$:

$$\frac{d}{da_0}\left(\sum_{i=1}^{n}(\ln v_i - \ln a_0 - t_i \ln(\alpha))^2\right) = \frac{2}{a_0}\sum_{i=1}^{n}(\ln a_0 - \ln v_i + t_i \ln(\alpha)) \stackrel{!}{=} 0,$$

hence $n \ln(a_0) + \ln(\alpha)\sum_{i=1}^{n} t_i - \sum_{i=1}^{n} \ln v_i = 0$ and

$$\frac{d}{d\alpha}\left(\sum_{i=1}^{n}(\ln v_i - \ln a_0 - t_i \ln(\alpha))^2\right) = \frac{2}{\alpha}\sum_{i=1}^{n} t_i (\ln a_0 - \ln v_i + t_i \ln(\alpha)) \stackrel{!}{=} 0.$$

So

$$\ln(a_0)\sum_{i=1}^{n} t_i + \ln(\alpha)\sum_{i=1}^{n} t_i^2 - \sum_{i=1}^{n} t_i \ln v_i = 0.$$

Approximating the logarithms has the intuitive effect of minimizing relative errors rather than absolute errors, because in the end we minimize the quadratic norm of

$$\varepsilon_i = \ln(a_0 \alpha^{t_i}) - \ln(v_i) = \ln\left(\frac{a_0 \alpha^{t_i}}{v_i}\right).$$

If the errors ε_i in the approximation are small enough, $|\varepsilon_i| < \delta$, we can use a first-order Taylor series:

$$1 - \delta \approx e^{-\delta} < \frac{a_0 \alpha^{t_i}}{v_i} < e^{\delta} \approx 1 + \delta \qquad\qquad \text{or}$$

$$v_i(1 - \delta) < a_0 \alpha^{t_i} < v_i(1 + \delta),$$

which is by definition an approximation of the relative error. This is valid as long as the approximation is good, i.e. for small δ.

2.3 Computer replacement and Moore's law

In 1965 Gordon Moore suggested that the density and number of components in integrated circuits was increasing exponentially and for its doubling time he stated "there is no reason to believe it will not remain nearly constant for at least 10 years."

Well, history proved him to be right much farther than expected. Nowadays his observation is usually called Moore's law and is modified (or approximated to reality) so that it states: every 18 months the ratio performance/price of computers doubles. This is clearly an exponential law, i.e. in 36 months the ratio will multiply by a factor of four, in 54 months by eight, etc. We will use this law as an axiom (a fact that we will not challenge) to derive the cost of a computer (and just for fun the cost of a floating point double-precision multiplication).

Let our units be days. Then the value of a computer which is t days old can be rewritten with Moore's law (assuming that a new computer will cost the same but will have improved performance) as:

$$v(t) = v_0 \left(\frac{1}{2}\right)^{t/(365 \cdot 1.5)}$$

since the performance of a computer does not change with time. The derivative of this value $v(t)$ with respect to t indicates how much value is lost per day:

$$v'(t) = -v_0 \frac{\ln(2)}{365 \cdot 1.5} \left(\frac{1}{2}\right)^{t/(365 \cdot 1.5)}$$

$$= -\frac{\ln(2)}{365 \cdot 1.5} \cdot v(t)$$

$$\approx -\frac{v(t)}{790}.$$

So a consequence of Moore's law is that a computer depreciates by $1/790$ of its value every day, i.e. if your computer costs \$7900 then the first day you lose \$10 and so on.

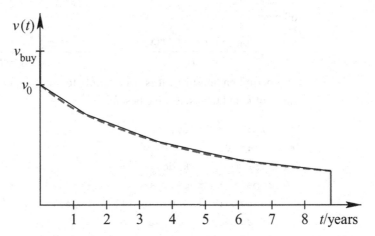

Figure 2.7 Exact model and piecewise linear approximation: dashes, exact model; line, piecewise linear approximation.

Notice that buying a computer for $7900 and keeping it idle for a month will cost you about $300!

A top of the line personal computer costs about $2000 and will perform 10^7 floating point multiplications per second. This means that a multiplication is worth:

$$\frac{\$2000}{790} \cdot \frac{1}{24} \cdot \frac{1}{3600} \cdot \frac{1}{10^7} \approx \$2.9 \cdot 10^{-12}.$$

That is about 3 picodollars (micro $= 10^{-6}$, nano $= 10^{-9}$, pico $= 10^{-12}$). The ratio of the entire GDP on planet Earth to $1 is as $1 to the cost of a multiplication!

2.3.1 Piecewise linear approximation

Real smooth functions are an abstraction of reality, very convenient for the mathematics, but sometimes not realistic. If we want to compute optimal values with more realistic data we have to use approximations to these ideal functions. A very important class of functions used for approximations are the piecewise linear functions.

In Section 2.2.3 we used the n given data points to approximate the exponential function. Here we use piecewise linear functions as approximations. (This method can be used, even when we do not have an exponential function.) In this case we use linear approximations between given data points (Figure 2.7). Piecewise linear approximations are very useful in practical situations, because of their simplicity and widespread applicability.

If we have data values at time $t_0 = 0$, t_1, t_2, ..., t_n, i.e. for $v(t)$ we have v_0, v_1, ..., v_n and for $r(t)$ we have r_0, r_1, ..., r_n, we can compute the costs per time as follows:

$$\frac{v_{\text{buy}} - v(t) + \int_0^{t_0} r(x)\mathrm{d}x}{t} \approx \frac{v_0 - v_k + \frac{1}{2}\sum_{i=1}^{k}(r_i + r_{i-1})(t_i - t_{i-1})}{t_k - t_0}$$

for a period ending in t_k. It is very simple to compute this sum for all the possible values of k and then select the best k:

```
lowest := infinity;
repair := 0;
for i from 1 to k do
    repair := repair + ( r[i] + r[i-1] ) * ( t[i]-t[i-1] ) /2;
    avcost := ( vbuy - v[i] + repair ) / t[i];
    if avcost < lowest then
        lowest := avcost;
        opttime := t[i]
    fi;
od;
print( "optimal time:", opttime );
print( "optimal average cost:", lowest );
```

3 Secondary structure prediction using least squares and singular value decomposition

Topics

- Secondary structure prediction (SSP)
- Modelling
- Least squares (LS)
- Singular value decomposition (SVD)
- Eigenvalue decomposition (EVD)

Secondary structure prediction in proteins is an essential problem of molecular biology.

See the web page of the European Molecular Biological Laboratory Heidelberg[1] and the ExPASy web pages[2] to get an impression of the multitude of tools used for SSP. We begin with a short summary of the biochemical facts.

3.1 Extremely brief introduction to protein biochemistry

BASIC

Proteins are basically chain molecules built up from sequences of amino acids (AAs). There are 20 different amino acids, as listed in Table 3.1. The chemical structures of the amino acids and further detailed information can be found, for example, in the Amino Acid Repository of the Jena Library of Biological Macromolecules[3] or at the FU Berlin.[4]

Protein structure can be described at three levels.

- The *primary structure* is the sequence of amino acids in the chain, i.e. a one-dimensional structure. The primary structures of millions of proteins are known

---

[1] Tools for SSP, EMBL: http://restools.sdsc.edu/biotools/biotools9.html
[2] Tools for SSP, ExPASy: www.expasy.org/tools/#secondary
[3] Jena Library: www.imb-jena.de/IMAGE_AA.html
[4] FU Berlin: www.chemie.fu-berlin.de/chemistry/bio/amino-acids_en.html

Table 3.1 Amino acids		
Full name	Three-letter abbreviation	One-letter abbreviation
Alanine	Ala	A
Arginine	Arg	R
Asparagine	Asn	N
Aspartic Acid	Asp	D
Cysteine	Cys	C
Glutamic Acid	Glu	E
Glutamine	Gln	Q
Glycine	Gly	G
Histidine	His	H
Isoleucine	Ile	I
Leucine	Leu	L
Lysine	Lys	K
Methionine	Met	M
Phenylalanine	Phe	F
Proline	Pro	P
Serine	Ser	S
Threonine	Thr	T
Tryptophan	Trp	W
Tyrosine	Tyr	Y
Valine	Val	V

today, many of them for several species. They can be found, among other places, in the SwissProt database.[5]

- The *secondary structure* is the result of the folding of parts of the AA-chain. The two most important secondary structures are the α-helix[6] and the β-sheet.[7] There exist other secondary structures which are of lesser importance such as turns, 3-10 helices, etc.

- The *tertiary structure* is the real three-dimensional configuration of the protein under given environmental conditions (solvent, pH and temperature). Two examples to be found on the internet are leghemoglobin,[8] a protein with a high content of α-helices, and concanavalin A,[9] a protein with high content of β-sheets.

The tertiary structure decides the biochemical function of the protein. If the tertiary structure is changed, the protein normally loses its ability to perform whatever function it has, since this function depends on the geometrical shape of the active site in the interior of the molecule (key–lock principle).

---

[5] SwissProt: www.expasy.ch/sprot/
[6] α-helix: http://en.wikipedia.org/wiki/Alpha_helix
[7] β-sheet: http://en.wikipedia.org/wiki/Beta_sheet
[8] leghemoglobin: http://ostracon.biologie.uni-kl.de/b_online/d17/17d.htm#09
[9] concanavalin A: http://ostracon.biologie.uni-kl.de/b_online/d17/cona.htm

Finding the primary structure is relatively easy, so millions of primary structures are known nowadays. Finding the secondary structure is difficult, and finding the tertiary structure is very difficult. Only about 50 000 tertiary structures are known today.[10] There are two main methods for finding three-dimensional structures experimentally: X-ray crystallography and nuclear magnetic resonance (NMR). Both methods require months of laboratory work for each structure. Finding the biochemical function of a protein is very difficult too.

Since the tertiary structure of a protein usually depends on interactions between amino acids which are far from each other in the primary structure, predicting these structures is a very difficult task. The prediction of secondary structures is slightly easier, since the interactions between neighboring amino acids play a larger role. In particular, the prediction of α-helices should be possible, since they depend mainly on the interactions between amino acids which are not more than four places in the chain away from each other. The reason for this lies in the fact, that an α-helix makes a $100°$ turn per amino acid. So 3.6 amino acids are a complete turn. The α-helices are stabilized by the interactions between next neighbors, i.e. between direct neighbors in the primary structure, and between amino acid numbers i and $i + 3$, and i and $i + 4$, which are neighbors in three-dimensional space.

3.1.1 The structure of myoglobin (*Physeter catodon*) – P02185

EXAMPLE

(i) Amino acid sequence and secondary structure notation[11]

    h:α-helix, t:turn, s:β-sheet or β-strand

AA number	2	5	10	15	20
Sequence	M V L S E G E W Q L V L H V W A K V E A D V A
Structure	h h h h h h h h h h h h h h h h h h h h

AA number	25	30	35	40	45
Sequence	G H G Q D I L I R L F K S H P E T L E K F D R
Structure	h h h h h h h h h h h h h h h h h h h t t

AA number		50	55	60	65
Sequence	F K H L K T E A E M K A S E D L K K H G V T V
Structure	t t t h h h h h h h h h h h h h h h

AA number	70	75	80	85	90
Sequence	L T A L G A I L K K K G H H E A E L K P L A Q
Structure	h h h h h h h h t t t t h h h h h h h h

---

[10] Number of tertiary structures: www.rcsb.org/
[11] See the Protein Data Bank (PDB) at www.pdb.org/pdb/explore/images.do?structureId =101M or slightly different structure data at http://www.expasy.org/cgi-bin/ protparam?P02185. Note the differences in the structures predicted from X-ray crystallography data by DSSP and Stride.

AA number	95	100	105	110	115

```
AA number        95        100        105        110        115
Sequence    S H A T K H K I P I K Y L E F I S E A I I H V
Structure   h h h h              h h h h h h h h h h h h h
```

```
AA number            120        125        130        135
Sequence    L H S R H P G D F G A D A Q G A M N K A L E L
Structure   h h h h     h h h         h h h h h h h h h h h
```

```
AA number    140        145        150        154
Sequence    F R K D I A A K Y K E L G Y Q G
Structure   h h h h h h h h h h h t t
```

(ii) Tertiary structure of this myoglobin molecule, see Figure 3.1.

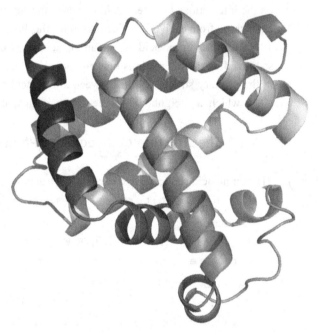

Figure 3.1 Tertiary structure of the myoglobin molecule of *Physeter catodon* (sperm whale).

3.1.2 Additional information about the biochemical structure of α-helices

BACKGROUND

(i) The formation of α-helices is a complicated process depending on many factors, not just the AA sequence, so predictions of their structure can only be approximate.

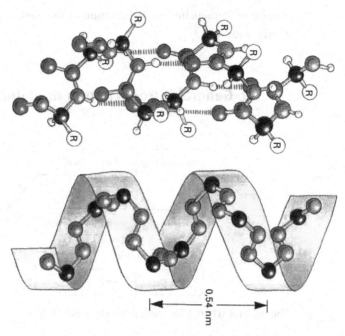

0.54 nm

Figure 3.2   Part of an α-helix, showing that 3.6 AA are necessary to complete one turn of the helix. (From B. Alberts, A. Johnson *et al.*, *Molecular Biology of the Cell*, 5th edition, Taylor and Francis, 2008, with kind permission of Taylor and Francis.)

(ii) In the primary structure the α-helices consist of sequences of approximately 6 to 20 amino acids.

(iii) α-helices are normally located near each other (often pairwise, parallel in a coiled coil structure[12]) in three dimensions.

(iv) An α-helix makes a 100° turn per amino acid (so 3.6 amino acids are a complete turn, see Figure 3.2).

(v) Typically there are "two sides" of an α-helix, one is hydrophobic ("water hating"), the other hydrophilic ("water loving"). So typically the hydrophobic parts will face the interior of the protein, the hydrophilic parts will face the outside, since proteins are normally surrounded by water.

(vi) On the average only 10% of the amino acids of a protein are part of some α-helix, so our input data will consist of sequences with batches of 6 to 20 "yes" interspersed between longer batches of "no." If we call $f(i)$ the function that will predict an α-helix at position i, then most of the time $f(i+1) = f(i)$. The actual percentage of amino acids in α-helices differs widely from protein to protein.

---

[12] Coiled coil: http://en.wikipedia.org/wiki/Coiled_coil

A concise introduction to the structure of the peptide bonds can be found on the World Wide Web.[13]

3.2 General introduction to modelling and prediction

EXAMPLE

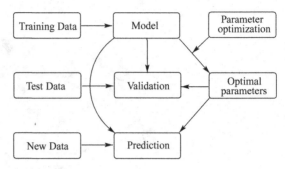

Figure 3.3   The general scheme for modelling and prediction.

Modelling and prediction can be described by Figure 3.3 which shows the main steps and components.

The first step is to use part of the existing data on known structures (called training data) to compute the best model parameters. Computing or approximating the parameters is normally called "training the model."

The second step is to use some other subset of the existing data (statistically independent of the training data) to validate the trained model. The result will usually be in the form of a percentage of correct predictions, or some other measure on the error/accuracy of the prediction.

Finally, if the validation is satisfactory, we are ready to make predictions on new data.

The box labelled "Model" in Figure 3.3 describes the mathematical function we use to make the predictions. The model, which we will call $f(\ldots)$, normally depends on values of the data and on some fixed parameters. The result of $f(\ldots)$ can be a boolean value (in decision problems, such as the prediction of α-helices) or a numerical value (in numerical problems, such as predicting the price of a stock, which is discussed in Chapter 7). Table 3.2 shows some example models.

In this section we shall be using models which are linear in both the data and the parameters.

---

[13] Structure of peptide bonds: www.med.unibs.it/~marchesi/aacids.html#peptide

Table 3.2 Example models (a_i are parameters, x_i are data values)
The last example is a computer program which can also be viewed as a model and as a mathematical function

$a_0 + a_1 x_1 + a_2 x_2$	numerical model, linear in both parameters and data
$a_0 + a_1 x_1 + a_2 x_2^2 > 0$	decision model, linear in parameters, quadratic in data
$a_1 x_1 + a_2 x_2 \approx \left\{ \begin{smallmatrix} 0 \\ 1 \end{smallmatrix} \right.$	decision model, linear in both parameters and data (the linear function can be other than 1 or 0)
$(x_1 < a_1$ and $x_1 \geq a_2)$ or $(\lvert x_2 \rvert > a_3$ and $x_1 < 0)$	non-linear decision model

```
proc(x1,x2)                                 non-linear numerical model
    for i to N do
        if Found(x1, t[i])
        then
            return(a0+a1*x2)
        fi
    od;
    return(a0+a1*(x1*x2))
end:
```

The box named "Parameter optimization" in Figure 3.3 describes the method that we will use to set the parameters to satisfy as much of the training data as possible. In this section we will use least squares. Examples of other techniques are linear programming and neural networks.

The boxes "Validation" and "Prediction" in Figure 3.3 are very similar and refer to computing $f(\ldots)$ on either test data (for validation) or new data (for prediction) using the optimized parameters.

We should make a precise division between traning data and test data. These two sets of data should contain independent sets of data, otherwise our validation is weak (or not existent). For example, humans and chimpanzees are 98% to 99% identical in their protein sequences. So including a human sequence in the training data and a chimp sequence in the test data is a useless and dangerous test because it gives a false sense of verification. In general, dividing the data into independent sets requires knowledge of the data.

Another problem arises when the test data are used more than once. If this is abused, i.e. we test one model with one set of test data after another to choose the model which validates best, we are de-facto using the test data as training data. A good rule to prevent this problem is to use the test data only once, i.e. optimize the model as much as you want, but validate it only once. Once the validation data are used, they can be used as training data, but not as validation data again.

The quality of the prediction is measured by the performance on the test data alone. Performance on the training data is only an indication when it is really poor. When the results of the training are poor, the model is probably worthless and there is no point in validating it.

Here is an overview of the various models and optimization techniques described here and in the following chapters.

Method	Mathematical structure	Optimized using	Chapter
Linear least squares	linear formula	SVD	3
Linear least squares	linear formula	best basis	4
Learning methods	k-dimensional search trees	nearest neighbors	5
Linear optimization	linear inequalities	simplex algorithm	6

3.2.1 SSP as modelling problem

Given a large set of examples of known helices and strands in proteins (e.g. 200 000), we shall use a subset of these as the *training set* and our goal is to predict, mathematically, the existence and position of α-helices in other proteins (in the *test set* or in totally new proteins).

EXAMPLE

$$
\begin{array}{ccccccccccc}
\cdots & R & A & A & D & T & G & G & S & D & P & \cdots \\
 & | & | & | & | & | & | & | & | & | & | & \\
\cdots & x & x & x & h & h & h & h & h & h & x & \cdots
\end{array}
$$

where h signifies a helix, x something else.

The α-helix prediction is a *decision problem*. We want to find a function $f(\ldots)$ which predicts, for example

$$
f(\text{AADTG}) = \begin{cases} \text{true} & \text{if} \quad \text{AADTG is part of a helix whenever it occurs} \\ \text{false} & \text{if} \quad \text{AADTG is never part of a helix.} \end{cases}
$$

Obviously we are in trouble when the sequence AADTG is part of a helix in one part of a protein, and part of a non-helix structure in another part. When something like this happens, the training data are inconsistent and our modelling cannot be perfect. Usually more information (a larger window of amino acids) removes inconsistencies.

A decision function, i.e. a function which returns *true* or *false*, can be implemented in many ways, for example for a numerical $f(\dots)$:

$$f(\dots) \geq d_0 \quad \Rightarrow \quad \text{Yes} - \text{it is an } \alpha\text{-helix,}$$
$$f(\dots) < d_0 \quad \Rightarrow \quad \text{No} - \text{it is not an } \alpha\text{-helix.}$$

Such a function can be constructed by training it to be one for the positives and zero for the negatives, i.e.

$$f(\dots) \approx 1 \quad \Rightarrow \quad \text{Yes} - \text{it is an } \alpha\text{-helix}$$
$$f(\dots) \approx 0 \quad \Rightarrow \quad \text{No} - \text{it is not an } \alpha\text{-helix,}$$

and determining the best splitting value d_0 after $f(\dots)$ is optimized in this way.

3.3 Least squares – singular value decomposition

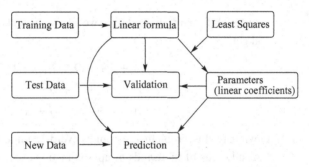

The scheme for modelling with least squares/singular value decomposition.

Using least squares with singular value decomposition (Figure 3.4), our predictor function f is constructed as follows. We cut the sequence of amino acids in pieces using a sliding window of width five. We shall not distinguish the positions of the amino acids inside the window, we only take into account how often an amino acid appears in it. We do this, since we want our function f to have "inertia," as the window moves. That is, the function at position i should not be very different from the function at position $i + 1$. This is motivated by the practical observation that helices are normally continuous blocks of 6 to 20 AAs long. Then a reasonable idea for

$$f(x) = a_A x_A + a_C x_C + a_D x_D + \cdots + a_Y x_Y$$

is to use as x_A, x_C, $\dots$ the number of occurrences of amino acids A, C, $\dots$ in the sliding window. These numbers change as we slide the window, but not by much, since it is only one AA out, one in, with each step.

We shall use the single-letter names of the AAs themselves instead of $a_A \ldots$, and we use #(X) for the number of amino acid X in the window, then we get the function definition:

$$f(x) = A \cdot \#(A) + C \cdot \#(C) + \cdots + Y \cdot \#(Y).$$

EXAMPLE   Here is a small part of an AA chain (primary structure): $\ldots$RAADTGGSDP$\ldots$
with known secondary structure $\ldots$xxxhhhhhhx$\ldots$
where h signifies an α-helix, x something else.

The value of the function should approximate whether the central amino acid in the sliding window is part of an α-helix ($f(x) = 1$) or not ($f(x) = 0$). Our simple model would result in the following approximations for the above example:

$$R + 2A + D + T \approx 0$$
$$2A + D + T + G \approx 1$$
$$A + D + T + 2G \approx 1 \quad \text{etc.} \tag{3.1}$$

To account for any constant bias, we add a constant c_0 to each of the above equations and get

$$R + 2A + D + T + c_0 \approx 0$$
$$2A + D + T + G + c_0 \approx 1$$
$$A + D + T + 2G + c_0 \approx 1 \quad \text{etc.} \tag{3.2}$$

Obviously this problem is strongly overdetermined! (We have 21 parameters and typically tens of thousands of approximations.) So we have to find an approximate solution. Least squares approximations are very common and useful, and will be the topic of this section.

Using the least squares approach we write the function to be minimized:

$$S^2 = (2A + D + R + T + c_0 - 0)^2 + (2A + D + G + T + c_0 - 1)^2$$
$$+ (A + D + 2G + T + c_0 - 1)^2 + \cdots.$$

Note that we have reordered the variables $A, C, D, \ldots, Y$. They are now in lexico-graphical order. S^2 measures the total error of the approximation. Our goal is to select the parameters $A, C, D, \ldots, Y$ and c_0 so that this error is minimized.

In matrix notation Equation (3.2) could be written as $\mathbf{A}x \approx b$, with

$$\mathbf{A} = \begin{pmatrix} 2 & 0 & 1 & \ldots & 1 \\ 2 & 0 & 1 & \ldots & 1 \\ 1 & 0 & 1 & \ldots & 1 \\ \vdots & \vdots & \vdots & \ldots & \vdots \end{pmatrix} \in \mathbb{R}^{m \times n}, \quad x = \begin{pmatrix} A \\ C \\ D \\ \vdots \\ Y \\ c_0 \end{pmatrix}, \quad b = \begin{pmatrix} 0 \\ 1 \\ 1 \\ \vdots \\ \end{pmatrix}$$

where m = number of rows = number of equations = number of training data points, n = number of columns = number of unknown parameters = number of AAs $+1 = 21$, and typically $m > 10\,000 \gg n$.

The matrix $\mathbf{A}$ contains in every row the information for one window, x is the vector of the unknown values to be computed (i.e. the names of the AAs in lexicographical order and c_0) and b contains one or zero depending on whether the central AA in the window is part of an α-helix or not.

We express our minimization problem with $\|r\| = \|\mathbf{A}x - b\|$, where our goal is to find the x which minimizes $\|r\|$. When this norm is an euclidean norm, the problem is a least squares problem.

See the interactive exercise "Modelling."

3.3.1 The need for weighting

Observations We have an imbalance of information: Since only about 10% of the AAs are in helices, we get fewer $f(\ldots) \approx 1$ than $f(\ldots) \approx 0$. Hence the trivial solution $f(\ldots) = 0$ is correct in 90% of cases, namely for all non-helices and for no helix. But $f(\ldots) = 0$ is not the solution we want! We want, for example, 80% correct non-helices and 80% correct helices.

To compensate for this imbalance of information, we have to give more weight to the relatively rare cases of helices. Therefore we introduce weights w_0 and w_1 where w_0 is the weight of making an error with $f(\ldots) = 0$ (false positive) and w_1 is the weight of an error for $f(\ldots) = 1$ (false negative). The weights have to be positive and be such that the total weight of all $f(\ldots) \approx 0$ cases is equal to the total weight of all $f(\ldots) \approx 1$ cases. It is also a good idea to normalize the weights, for example $w_0 + w_1 = 1$. Hence w_0 and w_1 are computed as follows:

$$w_0 = \frac{\#(f(\ldots) \approx 1)}{\#(\text{"1"s and "0"s})} = \frac{\|b\|^2}{m} \quad \text{and} \quad w_1 = \frac{\#(f(\ldots) \approx 0)}{\#(\text{"1"s and "0"s})} = \frac{m - \|b\|^2}{m}.$$

For example, when only 10% of AAs are in helices, we would have

$$w_0 = \frac{1}{10}, \quad w_1 = \frac{9}{10}, \quad \text{and} \quad w_0 \cdot \#(f(\ldots) \approx 0) = w_1 \cdot \#(f(\ldots) \approx 1).$$

With this kind of weighting applied to our example we get the following sum to minimize for the least squares approach:

$$S_2 = w_0(R + 2A + D + T + c_0 - 0)^2 + w_1(2A + D + T + G + c_0 - 1)^2$$
$$+ w_1(A + D + 2G + T + c_0 - 1)^2 + \cdots.$$

3.3.2 Singular value decomposition and eigenvalue decomposition

The singular value decomposition (SVD)

BASIC

The singular value decomposition (SVD) is one of the preferred methods for solving least squares problems. The singular values of a matrix $\mathbf{A} \in \mathbb{R}^{m \times n}$ are the values $\sigma_1 \geq \sigma_2 \geq \cdots \geq \sigma_n \geq 0$ defined by

$$\mathbf{A} = \mathbf{U} \mathbf{\Sigma} \mathbf{V}^{\mathrm{T}},$$

where $\mathbf{U}$ and $\mathbf{V}$ are orthonormal $m \times m$ respectively $n \times n$ matrices, i.e. $\mathbf{U}\mathbf{U}^{\mathrm{T}} = \mathbf{I} \in \mathbb{R}^{m \times m}$, $\mathbf{V}\mathbf{V}^{\mathrm{T}} = \mathbf{I} \in \mathbb{R}^{n \times n}$, which implies $\|\mathbf{U}\| = \|\mathbf{V}\| = 1$, and where $\mathbf{\Sigma}$ is a diagonal $m \times n$ matrix with the singular values σ_i in the diagonal (see below).

There exists an efficient method for computing these values σ_i and the transformation matrix $\mathbf{V}$, which is called the SVD (singular value decomposition). Normally we do not want to compute $\mathbf{U}$, which is very large ($m \times m$).

We will now show how to use SVD for our problem of least squares:

$$\|\mathbf{r}\| := \|\mathbf{A}\mathbf{x} - \mathbf{b}\| = \left\|\mathbf{U}\mathbf{\Sigma}\mathbf{V}^{\mathrm{T}}\mathbf{x} - \mathbf{b}\right\| = \left\|\mathbf{\Sigma}\underbrace{\mathbf{V}^{\mathrm{T}}\mathbf{x}}_{z} - \underbrace{\mathbf{U}^{\mathrm{T}}\mathbf{b}}_{c}\right\|.$$

The last equality is true, since we have multiplied by $\mathbf{U}^{-1} = \mathbf{U}^{\mathrm{T}}$, and $\left\|\mathbf{U}^{\mathrm{T}}\right\| = \|\mathbf{U}\| = 1$, since $\mathbf{U}$ is orthonormal.

So minimizing $\|\mathbf{r}\|$ is equivalent to minimizing $\|\mathbf{\Sigma} \cdot \mathbf{z} - \mathbf{c}\|$. Written out in components, this means minimizing

$$\|r\| = \left\| \begin{pmatrix} \sigma_1 & & 0 \\ & \ddots & \\ 0 & & \sigma_n \\ 0 & \cdots & 0 \\ \vdots & \ddots & \vdots \\ 0 & \cdots & 0 \end{pmatrix} \cdot \begin{pmatrix} z_1 \\ \vdots \\ z_n \end{pmatrix} - \begin{pmatrix} c_1 \\ \vdots \\ c_n \\ c_{n+1} \\ \vdots \\ c_m \end{pmatrix} \right\| \quad \text{with} \quad z = \mathbf{V}^{\mathrm{T}}x \quad \text{or} \quad x = \mathbf{V}z.$$

We now choose z_k so that $\|\mathbf{r}\|$ is minimal. We can make all the components of $\mathbf{r}$ which have $\sigma_i \neq 0$ for $k = 1, \ldots, n$ equal to 0. The rest, from r_{n+1} to r_m, cannot be changed and determines the minimal size of $\|\mathbf{r}\|$:

$$z_k = \frac{c_k}{\sigma_k} \quad \text{for} \quad 1 \leq k \leq n, \quad \text{when } \sigma_k > 0$$

$$\text{and} \quad z_k = 0 \quad \text{for} \quad 1 \leq k \leq n, \quad \text{when} \quad \sigma_k = 0. \tag{3.3}$$

The actual calculation of the SVD

A desirable improvement to minimize the number of calculations For our kind of problem we have $\mathbf{A} \in \mathbb{R}^{m \times n}$ with $m > 10\,000$ and $n = 21$. To avoid matrices of size $10\,000 \times 21$, it would be nice if we could calculate everything using only $\mathbf{A}^T\mathbf{A} \in \mathbb{R}^{n \times n}$, $\mathbf{b}^T\mathbf{A} \in \mathbb{R}^{1 \times n}$, and $\mathbf{b}^T\mathbf{b} \in \mathbb{R}$.

$\mathbf{A}^T\mathbf{A}$, $\mathbf{b}^T\mathbf{A}$ and $\mathbf{b}^T\mathbf{b}$ can be calculated easily and efficiently when moving the sliding window across the sequences. This mode of computing is called "online," as each data point analyzed is processed completely and does not need to be stored for later use. This has two advantages: it not only saves us from calculating huge matrices with dimension $m \times m$, but it also allows us to compute an approximation, and then add more data to compute a new one.

For example, let $RAADT$ be a window of AAs at position i (i.e. the midpoint of the window is at position i). Let $\mathbf{A}_i$ be the vector such that

$$\mathbf{A}_i \cdot \mathbf{x} = R + A + A + D + T + c_0,$$

i.e.

$$\mathbf{A}_i = (2, 0, 1, \ldots, 0, 1).$$

In this case $\mathbf{A}_i$ is the ith row of the matrix $\mathbf{A}$. Let $b_i = 1$ if the AA at position i is part of an α-helix, $b_i = 0$ if the AA at position i is not part of an α-helix.

Starting with a zero matrix $\mathbf{A}^T\mathbf{A} = \mathbf{0}$, a zero vector $\mathbf{b}^T\mathbf{A} = \mathbf{0}$ and $\mathbf{b}^T\mathbf{b} = 0$ we compute for $i = 1$ to m:

$$\mathbf{A}^T\mathbf{A} + \mathbf{A}_i^T \cdot \mathbf{A}_i \longrightarrow \mathbf{A}^T\mathbf{A}$$

$$\mathbf{b}^T\mathbf{A} + b_i \cdot \mathbf{A}_i^T \longrightarrow \mathbf{b}^T\mathbf{A}$$

$$\mathbf{b}^T\mathbf{b} + b_i^2 \longrightarrow \mathbf{b}^T\mathbf{b}.$$

These equations hold, since:

$$\mathbf{A}^T\mathbf{A} = \begin{pmatrix} a_{11} & a_{21} & \cdots & a_{m1} \\ a_{12} & a_{22} & \cdots & a_{m2} \\ a_{13} & a_{23} & \cdots & a_{m3} \\ \vdots & \vdots & \ddots & \vdots \\ a_{1n} & a_{2n} & \cdots & a_{mn} \end{pmatrix} \cdot \begin{pmatrix} a_{11} & a_{12} & a_{13} & \cdots & a_{1n} \\ a_{21} & a_{22} & a_{23} & \cdots & a_{2n} \\ \vdots & \vdots & \vdots & \ddots & \vdots \\ a_{m1} & a_{m2} & a_{m3} & \cdots & a_{mn} \end{pmatrix}$$

$$= \begin{pmatrix} a_{11}^2 + a_{21}^2 + \cdots + a_{m1}^2 & \cdots & a_{11}a_{1n} + \cdots + a_{m1}a_{mn} \\ a_{12}a_{11} + \cdots + a_{m2}a_{m1} & \cdots & a_{12}a_{1n} + \cdots + a_{m2}a_{mn} \\ a_{13}a_{11} + \cdots + a_{m3}a_{m1} & \cdots & a_{13}a_{1n} + \cdots + a_{m3}a_{mn} \\ \vdots & \ddots & \vdots \\ a_{1n}a_{11} + \cdots + a_{mn}a_{m1} & \cdots & a_{1n}^2 + a_{2n}^2 + \cdots + a_{mn}^2 \end{pmatrix} \qquad (3.4)$$

and

$$\sum_{i=1}^{m} \mathbf{A}_i^{\mathrm{T}} \mathbf{A}_i = \sum_{i=1}^{m} \begin{pmatrix} a_{i1} \\ a_{i2} \\ \vdots \\ a_{in} \end{pmatrix} \cdot \begin{pmatrix} a_{i1} & a_{i2} & \cdots & a_{in} \end{pmatrix}.$$

Note that this is the exterior product of the two vectors, resulting in a matrix of dimension $n \times n$.

So

$$\sum_{i=1}^{m} \mathbf{A}_i^{\mathrm{T}} \mathbf{A}_i = \sum_{i=1}^{m} \begin{pmatrix} a_{i1}^2 & a_{i1}a_{i2} & \cdots & a_{i1}a_{in} \\ a_{i2}a_{i1} & a_{i2}^2 & \cdots & a_{i2}a_{in} \\ \vdots & \vdots & \ddots & \vdots \\ a_{in}a_{i1} & a_{in}a_{i2} & \cdots & a_{in}^2 \end{pmatrix}$$

$$= \begin{pmatrix} a_{11}^2 + a_{21}^2 + \cdots + a_{m1}^2 & \cdots & a_{11}a_{1n} + \cdots + a_{m1}a_{mn} \\ a_{12}a_{11} + \cdots + a_{m2}a_{m1} & \cdots & a_{12}a_{1n} + \cdots + a_{m2}a_{mn} \\ \vdots & \ddots & \vdots \\ a_{1n}a_{11} + \cdots + a_{mn}a_{m1} & \cdots & a_{1n}^2 + a_{2n}^2 + \cdots + a_{mn}^2 \end{pmatrix}.$$

Hence $\sum_{i=1}^{m} \mathbf{A}_i^{\mathrm{T}} \mathbf{A}_i = \mathbf{A}^{\mathrm{T}} \mathbf{A}$.

Similarly

$$\boldsymbol{b}^{\mathrm{T}} \mathbf{A} = \begin{pmatrix} b_1 & b_2 & \cdots & b_m \end{pmatrix} \cdot \begin{pmatrix} a_{11} & a_{12} & a_{13} & \cdots & a_{1n} \\ a_{21} & a_{22} & a_{23} & \cdots & a_{2n} \\ \vdots & \vdots & \vdots & \ddots & \vdots \\ a_{m1} & a_{m2} & a_{m3} & \cdots & a_{mn} \end{pmatrix}$$

$$= \begin{pmatrix} b_1 a_{11} + b_2 a_{21} + \cdots + b_m a_{m1} \\ b_1 a_{12} + b_2 a_{22} + \cdots + b_m a_{m2} \\ b_1 a_{13} + b_2 a_{23} + \cdots + b_m a_{m3} \\ \vdots \\ b_1 a_{1n} + b_2 a_{2n} + \cdots + b_m a_{mn} \end{pmatrix}$$

can be calculated as

$$\sum_{i=1}^{m} b_i \mathbf{A}_i^{\mathrm{T}} = \sum_{i=1}^{m} b_i \begin{pmatrix} a_{i1} \\ a_{i2} \\ \vdots \\ a_{in} \end{pmatrix} = \sum_{i=1}^{m} \begin{pmatrix} b_i a_{i1} \\ b_i a_{i2} \\ \vdots \\ b_i a_{in} \end{pmatrix}$$

and

$$b^{\mathrm{T}}b = \begin{pmatrix} b_1 & b_2 & \cdots & b_m \end{pmatrix} \cdot \begin{pmatrix} b_1 \\ b_2 \\ \vdots \\ b_m \end{pmatrix} = b_1^2 + b_2^2 + \cdots + b_m^2 = \sum_{i=1}^{m} b_i^2.$$

In the next subsection we shall show how to compute the least squares solution using only $A^{\mathrm{T}}A$, $b^{\mathrm{T}}A$ and $b^{\mathrm{T}}b$ (without using A or b).

Least squares using eigenvalue/eigenvector decomposition

We can express $\|r\|^2$ in the following way:

$$\begin{aligned}
\|r\|^2 &= \|Ax - b\|^2 \\
&= (Ax - b)^{\mathrm{T}}(Ax - b) \\
&= (x^{\mathrm{T}}A^{\mathrm{T}} - b^{\mathrm{T}})(Ax - b) \\
&= x^{\mathrm{T}}A^{\mathrm{T}}Ax - x^{\mathrm{T}}A^{\mathrm{T}}b - b^{\mathrm{T}}Ax + b^{\mathrm{T}}b \\
&= x^{\mathrm{T}}A^{\mathrm{T}}Ax - 2x^{\mathrm{T}}A^{\mathrm{T}}b + b^{\mathrm{T}}b,
\end{aligned}$$

since $x^{\mathrm{T}}A^{\mathrm{T}}b = b^{\mathrm{T}}Ax \in \mathbb{R}$.

Notice from Equation (3.4) that $A^{\mathrm{T}}A$ is symmetric and will therefore have real eigenvalues. Let its eigenvalue decomposition be

$$A^{\mathrm{T}}A = W\Lambda W^{\mathrm{T}},$$

where W is an orthonormal $n \times n$ matrix, and Λ is a diagonal matrix with diagonal entries (eigenvalues) λ_i and $\lambda_1 \geq \lambda_2 \geq \cdots \geq \lambda_n$. And we get

$$\|r\|^2 = x^{\mathrm{T}}W\Lambda W^{\mathrm{T}}x - 2x^{\mathrm{T}}A^{\mathrm{T}}b + b^{\mathrm{T}}b.$$

Applying the definition $z := W^{\mathrm{T}}x$ we get $x = Wz$ since $W^{\mathrm{T}} = W^{-1}$ and:

$$\begin{aligned}
\|r\|^2 &= z^{\mathrm{T}}\Lambda z - 2(z^{\mathrm{T}}W^{\mathrm{T}})A^{\mathrm{T}}b + b^{\mathrm{T}}b \\
&= z^{\mathrm{T}}\Lambda z - 2z^{\mathrm{T}}d + b^{\mathrm{T}}b
\end{aligned}$$

where $d := W^{\mathrm{T}}A^{\mathrm{T}}b$. Then, since W^{T} is $n \times n$ and $A^{\mathrm{T}}b$ is a vector of dimension n it follows that d is a vector with dimension n.

Since $\mathbf{A}$ is diagonal, the product $z^{\mathrm{T}}\mathbf{A}z$ is very simple and is the sum of the diagonal elements times the squares of the z_i:

$$\|r\|^2 = z^{\mathrm{T}}\mathbf{A}z - 2z^{\mathrm{T}}d + b^{\mathrm{T}}b$$

$$= \sum_{i=1}^{n} \left(\lambda_i z_i^2 - 2d_i\, z_i\right) + b^{\mathrm{T}}b.$$

Each term of this sum depends only on one value z_i of z.

The minimum of this equation is achieved by the minimum of each quadratic term (minimized with respect to z_i), i.e. the derivative of each term with respect to z_i equated to 0:

$$2\lambda_i z_i - 2d_i = 0 \implies z_i = \frac{d_i}{\lambda_i}.$$

The relation between the singular value decomposition of $\mathbf{A}$ and the eigenvalue decomposition of $\mathbf{A}^{\mathrm{T}}\mathbf{A}$

ADVANCED

Let $\mathbf{A} = \mathbf{U}\mathbf{\Sigma}\mathbf{V}^{\mathrm{T}}$ be a SVD of $\mathbf{A}$. Then $\mathbf{A}^{\mathrm{T}}\mathbf{A} = \mathbf{V}\mathbf{\Sigma}^{\mathrm{T}}\mathbf{U}^{\mathrm{T}}\mathbf{U}\mathbf{\Sigma}\mathbf{V}^{\mathrm{T}} = \mathbf{V}\mathbf{\Sigma}^{\mathrm{T}}\mathbf{\Sigma}\mathbf{V}^{\mathrm{T}}$.

On the other hand, from the eigenvalue decomposition, $\mathbf{A}^{\mathrm{T}}\mathbf{A} = \mathbf{W}\mathbf{\Lambda}\mathbf{W}^{\mathrm{T}}$, since $\mathbf{A}^{\mathrm{T}}\mathbf{A}$ is symmetric, so

$$\mathbf{W}\mathbf{\Lambda}\mathbf{W}^{\mathrm{T}} = \mathbf{V}\mathbf{\Sigma}^{\mathrm{T}}\mathbf{\Sigma}\mathbf{V}^{\mathrm{T}},$$

since $\mathbf{A}^{\mathrm{T}}\mathbf{A}$ has (up to ordering) a unique eigenvalue decomposition. Since $\mathbf{\Sigma}^{\mathrm{T}}\mathbf{\Sigma}$ is a diagonal matrix, with diagonal entries σ_i^2, it is obvious that $\mathbf{\Sigma}^{\mathrm{T}}\mathbf{\Sigma} = \mathbf{\Lambda}$, and $\mathbf{W} = \mathbf{V}$.

This is true except for the possible variations in the order of the eigenvalues and singular values. Since we have chosen a decreasing order for both, the only remaining ambiguity results from equal eigenvalues, which is inconsequential for our analysis.

In other words: $\sigma_i = \sqrt{\lambda_i}$, and $(\mathbf{V} = \mathbf{W}, \mathbf{\Sigma}^2 = \mathbf{\Lambda})$ is the eigenvector/eigenvalue decomposition of $\mathbf{A}^{\mathrm{T}}\mathbf{A}$.

Remark: $\mathbf{A}\mathbf{A}^{\mathrm{T}} = \mathbf{U}\mathbf{\Sigma}\mathbf{V}^{\mathrm{T}}\,\mathbf{V}\mathbf{\Sigma}^{\mathrm{T}}\mathbf{U}^{\mathrm{T}} = \mathbf{U}\mathbf{\Sigma}\mathbf{\Sigma}^{\mathrm{T}}\mathbf{U}^{\mathrm{T}}$, so $(\mathbf{U}, \mathbf{\Lambda})$ is the eigenvector/ eigenvalue decomposition of $\mathbf{A}\mathbf{A}^{\mathrm{T}}$. This means that if we need $\mathbf{U}$ we would have to compute $\mathbf{A}\mathbf{A}^{\mathrm{T}}$ which is of dimension $m \times m$, i.e. usually very large. Fortunately $\mathbf{U}$ is not needed in our applications.

For more details try the interactive exercise "SVD."

3.3.3 Criteria for discarding singular values

The main advantage of using SVD (or eigenvalue decomposition) lies in the ability to deal with singular or almost singular problems. This ability comes from being able to discard non-significant singular values and allows us to solve three important problems.

- **Singular problems** These may be caused by a linear dependence in the variables which we did not notice. It may also arise casually with independent variables which take discrete values, e.g. x_1 is either zero or one, x_2 is either two or three, and in our data we only have occurrences of $(x_1 = 0, x_2 = 3)$ and $(x_1 = 1, x_2 = 2)$. Singular problems will have one σ_i equal to zero for each linear dependence. The number of σ_i equal to zero is the dimension of the null space of the matrix $\mathbf{A}^\mathsf{T}\mathbf{A}$.

- **Nearly singular problems** These problems are singular problems as above, which due to rounding errors in the data or in the computation are technically not singular. Nearly singular problems will have σ_i which are of the order of the rounding errors, i.e. very small compared to the others. For example, if $\sigma_n < \varepsilon \sigma_1$, where ε is the machine epsilon, then σ_n is probably just roundoff error.

- **Insignificant contributions** These arise normally by ill-conditioned problems with peculiar distributions in the independent variables, which are not completely independent of each other. When these solutions are included we have a large increase in the norm $\|x\|$ of the solution at the benefit of a very small reduction in the norm of the residuals $\|r\|$.

The solution of these three problems is the same: discard the smallest singular values. We need to decide the cutoff point k, i.e. we will use the singular values $\sigma_1, \sigma_2, \dots, \sigma_k$ and discard the rest, so we set $z_{k+1}, \dots, z_n = 0$.

Using all the singular values, we have (see Equation (3.3))

$$\|r\|^2 = c_{n+1}^2 + \dots + c_m^2 \quad \text{and} \quad \|x\|^2 = \|z\|^2 = \sum_{k=1}^{n} \left(\frac{c_k}{\sigma_k} \right)^2.$$

Now, what happens when we neglect a small singular value σ_n by setting $z_n = 0$?

$$\|\tilde{x}\|^2 = \|\tilde{z}\|^2 = \sum_{k=1}^{n-1} \left(\frac{c_k}{\sigma_k} \right)^2 \quad \text{and} \quad \|\tilde{r}\|^2 = c_n^2 + c_{n+1}^2 + \dots + c_m^2.$$

Comparing the decrease in the norm of the solution vector x with the increase in the norm of the error r, we get

$$\frac{\Delta \|x\|^2}{\Delta \|r\|^2} = \frac{\|x\|^2 - \|\tilde{x}\|^2}{\|\tilde{r}\|^2 - \|r\|^2} = \frac{(c_n/\sigma_n)^2}{c_n^2} = \frac{1}{\sigma_n^2} = \frac{1}{\lambda_n}.$$

When σ_k is small, the omission of $z_k = c_k/\sigma_k$ will decrease $\|x\|$ significantly while increasing $\|r\|$ by little. Hence it may be desirable to ignore the contributions of z_k

for very small σ_k; the omission of such values will be justified by the large decrease in $\|x\|$ and the small increase in $\|r\|^2$ by c_k^2 (see Figures 3.5 and 3.6).

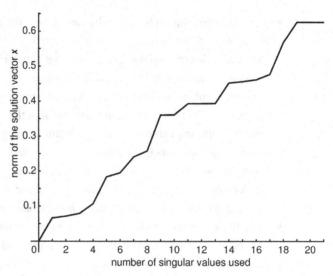

Figure 3.5 $\|\tilde{x}\|^2$ **versus the number of singular values used (see Table 3.3).**

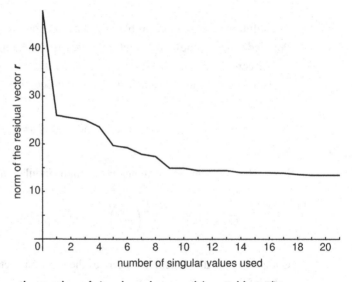

Figure 3.6 $\|\tilde{r}\|^2$ **versus the number of singular values used (see Table 3.3).**

PRACTICAL NOTE   Why do we care about $\|x\|$ being very low? This is a very valid question, since it is reasonable to assume that what matters is the norm of the error $\|r\|$ more than anything else. It happens that for problems which include variables which have little or no influence (and hence in principle should be ignored) their coefficients have to become very large to minimize $\|r\|$. Normally these large coefficients will be of opposite signs, which has the effect of cancelling the possible signal and magnifying the noise. It is for this reason that we want to be careful about how much increase in $\|x\|$ we will have for a decrease in $\|r\|$.

So, to simplify the problem we only want to take into account the k largest singular values, namely $\sigma_1, \sigma_2, \ldots, \sigma_k$, and we want to discard $\sigma_{k+1}, \ldots, \sigma_n$. (Note: $\sigma_{i+1} \leq \sigma_i$.) So how should we choose k?

Criteria for choosing k

(i) $\sigma_{k+1} = 0$. (The data have $n - k$ dependent columns.)

(ii) $\sigma_{k+1} < 0 +$ numerical roundoff error. (Same as above; in practice roundoff errors will prevent the singular values from being exactly zero.)

(iii) $\sigma_{k+1}/\sigma_1 < 1/1000$. (Contribution of variable $k + 1$ is 1000 times smaller than the biggest contribution.)

(iv) $1/\sigma_{k+1}^2$ is too large, based on some knowledge of what $\|r\|$ and $\|x\|$ represent.

3.3.4 A concrete example

EXAMPLE

As an example we apply our mathematical model to the AA sequence of the cytochrome c_2 precursor, which consists of 112 AAs of which 48 are part of helices. Notation: h α-helix, t turn, s β-sheet

```
Sequence    E G D A A A G E K V S K K C L A C H T F D Q G
Structure       h h h h h h h h h h h t t t       s t t

Sequence    G A N K V G P N L F G V F E N T A A H K D N Y
Structure         s s s       t t   t t s s s s     t t s
Sequence    A Y S E S Y T E M K A K G L T W T E A N L A A
Structure       h h h h h h h h t t   s     h h h h h h

Sequence    Y V K N P K A F V L E K S G D P K A K S K M T
Structure   h h h   h h h h h h h h     t t       s

Sequence    F K L T K D D E I E N V I A Y L K T L K
Structure         h h h h h h h h h h h t t
```

The matrix $A^T A$ computed for this sequence of the cytochrome c_2 precursor is:

$$A^T A = \begin{bmatrix}
118 & 0 & 27 & 16 & 7 & 2 & 22 & 26 & 10 & 4 & 21 & 43 & 4 & 6 & 6 & 7 & 16 & 2 & 19 & 14 & 74 \\
0 & 0 & 0 & 0 & 0 & 0 & 0 & 0 & 0 & 0 & 0 & 0 & 0 & 0 & 0 & 0 & 0 & 0 & 0 & 0 & 0 \\
27 & 0 & 35 & 5 & 0 & 1 & 13 & 13 & 3 & 6 & 8 & 15 & 0 & 7 & 9 & 1 & 6 & 1 & 10 & 15 & 35 \\
16 & 0 & 5 & 36 & 1 & 4 & 12 & 12 & 5 & 5 & 3 & 18 & 0 & 4 & 4 & 3 & 8 & 0 & 4 & 0 & 28 \\
7 & 0 & 0 & 1 & 14 & 0 & 0 & 0 & 5 & 0 & 7 & 8 & 0 & 2 & 0 & 2 & 3 & 0 & 0 & 1 & 10 \\
2 & 0 & 1 & 4 & 0 & 5 & 0 & 7 & 1 & 0 & 0 & 0 & 0 & 3 & 0 & 0 & 2 & 0 & 0 & 0 & 5 \\
22 & 0 & 13 & 12 & 0 & 0 & 49 & 9 & 0 & 10 & 7 & 15 & 4 & 7 & 0 & 15 & 16 & 3 & 10 & 13 & 41 \\
26 & 0 & 13 & 12 & 0 & 7 & 9 & 45 & 0 & 0 & 10 & 20 & 1 & 11 & 8 & 5 & 5 & 2 & 0 & 11 & 37 \\
10 & 0 & 3 & 5 & 5 & 1 & 0 & 0 & 10 & 0 & 2 & 4 & 0 & 3 & 0 & 0 & 6 & 0 & 1 & 0 & 10 \\
4 & 0 & 6 & 5 & 0 & 0 & 10 & 0 & 0 & 12 & 2 & 2 & 0 & 0 & 0 & 0 & 0 & 0 & 3 & 6 & 10 \\
21 & 0 & 8 & 3 & 7 & 0 & 7 & 10 & 2 & 2 & 41 & 28 & 1 & 11 & 3 & 3 & 19 & 3 & 7 & 9 & 37 \\
43 & 0 & 15 & 18 & 8 & 0 & 15 & 20 & 4 & 2 & 28 & 126 & 14 & 11 & 16 & 25 & 23 & 1 & 9 & 22 & 80 \\
4 & 0 & 0 & 0 & 0 & 0 & 4 & 1 & 0 & 0 & 1 & 14 & 10 & 3 & 0 & 4 & 7 & 0 & 2 & 0 & 10 \\
6 & 0 & 7 & 4 & 2 & 3 & 7 & 11 & 3 & 0 & 11 & 11 & 3 & 29 & 4 & 1 & 12 & 0 & 0 & 11 & 25 \\
6 & 0 & 9 & 4 & 0 & 0 & 0 & 8 & 0 & 0 & 3 & 16 & 0 & 4 & 15 & 3 & 0 & 0 & 1 & 6 & 15 \\
7 & 0 & 1 & 3 & 2 & 0 & 15 & 5 & 0 & 0 & 3 & 25 & 4 & 1 & 3 & 31 & 6 & 0 & 14 & 5 & 25 \\
16 & 0 & 6 & 8 & 3 & 2 & 16 & 5 & 6 & 0 & 19 & 23 & 7 & 12 & 0 & 6 & 46 & 8 & 6 & 1 & 38 \\
2 & 0 & 1 & 0 & 0 & 0 & 3 & 2 & 0 & 0 & 3 & 1 & 0 & 0 & 0 & 0 & 8 & 5 & 0 & 0 & 5 \\
19 & 0 & 10 & 4 & 0 & 0 & 10 & 0 & 1 & 3 & 7 & 9 & 2 & 0 & 1 & 14 & 6 & 0 & 33 & 6 & 25 \\
14 & 0 & 15 & 0 & 1 & 0 & 13 & 11 & 0 & 6 & 9 & 22 & 0 & 11 & 6 & 5 & 1 & 0 & 6 & 30 & 30 \\
74 & 0 & 35 & 28 & 10 & 5 & 41 & 37 & 10 & 10 & 37 & 80 & 10 & 25 & 15 & 25 & 38 & 5 & 25 & 30 & 108
\end{bmatrix}$$

$b^T A = [40, 0, 14, 10, 3, 0, 31, 9, 0, 10, 15, 36, 5, 5, 4, 14, 9, 1, 15, 19, 48]$

$b^T b = 48.$

The result of the singular value decomposition of these data (in order of decreasing absolute values of the coefficients) is as follows.

```
108 data points,14 21 independent variables
19 singular values used:   2.515   3.077   4.59   5.117   7.129
   7.644   10.56   12.3   15.44   21.13   23.7   28.01   29.93   41.04
   50.47   51.3   64.29   87.69   332.1
2 singular values discarded: -2.26e-15   0
```

[14] For a window of size 5, there are two positions at each end which cannot be used.

variable	coeff	variable	coeff	
Glu	0.3758	Leu	0.11636	
His	-0.27822	Phe	-0.11255	norm of data points:
Gln	0.26416	Tyr	0.064367	48
Val	0.25756	Lys	0.059713	
Gly	-0.21072	Pro	0.046629	norm of residuals:
Ile	0.20583	Trp	0.038181	13.451
Ala	0.18265	Cys	0.037322	
c0	0.17956	Ser	0.0016823	norm when all SV used:
Thr	-0.17835	Asp	0.0014112	13.451
Met	0.16796	Arg	0	
Asn	-0.14197			

The first thing to notice is that our problem had some linear dependences (due to very little data for some amino acids). One singular value is zero and the next one is within roundoff error. These two are discarded as discussed earlier.

Interpretation of results

Recall that $f(\ldots)$ is the sum of the values of the amino acids in a window of size 5 around the amino acid in question. This sum should be equal to one if the middle position is a helix or zero if it is not. So positive values should be viewed as helix-forming and negative values as helix-breaking.

So we can conclude that Glu (0.3758), Gln (0.26416) and Val (0.25756), having the largest positive values, are helix-forming AAs, and His (–0.27822), Gly (–0.21072) and Thr (–0.17835), having negative values, are helix-breaking AAs.

Remark Note that this example is based on too little data to allow serious conclusions. More solid conclusions would be obtained from hundreds or thousands of proteins.

What happens if we discard some singular values? Table 3.3 shows that the norm of the error changes very little if we use 14 singular values instead of 19. This can also be seen in a graphical way in Figure 3.6, where the norm of the error, $\|\tilde{r}^2\|$, remains flat above $k = 14$. Other interesting cutoff points are between 4 and 9 singular values. At these points we have either a decrease in the norm of the residual error or an increase in the norm of the solution vector or both. This can be concluded from the values in Table 3.3 and from the graphs in Figures 3.5 and 3.6 that show the reduction in the error and the increase in the norm of the solution vector.

Table 3.3 The size of the norm $\|\tilde{x}^2\|$ of the solution and of the norm $\|\tilde{r}^2\|$ of the residue depending on the number of the singular values

Number of singular values used	Singular value	$\|\tilde{x}^2\|$	$\|\tilde{r}^2\|$
0		0.0000	48.0000
1	332.0659	0.0663	25.9679
2	87.6890	0.0716	25.5057
3	64.2914	0.0795	25.0005
4	51.2999	0.1069	23.5935
5	50.4718	0.1840	19.7016
6	41.0402	0.1956	19.2245
7	29.9331	0.2409	17.8686
8	28.0126	0.2578	17.3975
9	23.6977	0.3605	14.9632
10	21.1295	0.3607	14.9591
11	15.4379	0.3932	14.4567
12	12.2969	0.3935	14.4535
13	10.5622	0.3937	14.4514
14	7.6440	0.4523	14.0031
15	7.1290	0.4563	13.9750
16	5.1167	0.4615	13.9480
17	4.5902	0.4769	13.8775
18	3.0766	0.5683	13.5964
19	2.5154	0.6262	13.4505
20	0.0000	0.6262	13.4505
21	−0.0000	0.6262	13.4505

The largest of the singular values, σ_1, has a value of 332, so the variable $\mathbf{V}_1$ (the first eigenvector of the transformation matrix $\mathbf{V}$) is the most important linear combination of variables. Likewise all the large components of $\mathbf{V}_1$ correspond to important variables in our original basis. The same holds for all other large σ_i.

Using the nine largest eigenvalues, we get the following results.

```
Results of SVD analysis ordered with decreasing absolute values
108 data points, 21 independent variables
9 singular values used:  23.7  28.01  29.93  41.04  50.47  51.3
      64.29  87.69  332.1
12 singular values discarded:  -2.26e-15  0  2.52  3.08  4.59
      5.12  7.13  7.64  10.6  12.3  15.44  21.13
```

variable	coeff	variable	coeff		
Glu	0.39938	Pro	-0.079242		
Val	0.18728	Gln	-0.069081		
Ile	0.16377	His	-0.056838	Norm of raw data	
Thr	-0.16052	Cys	-0.048291	48	
Ala	0.15724	Met	-0.029052	norm of residuals	
c0	0.12493	Trp	-0.023787	14.963	
Leu	0.11684	Tyr	0.022116	norm when all SV used	
Gly	-0.11569	Asn	-0.019377	13.451	
Lys	0.096441	Asp	0.010231		
Phe	-0.095369	Arg	0		
Ser	0.092501				

And using only four eigenvalues we get the following results.

```
Results of SVD analysis, 108 data points, 21 independent variables
4 singular values used:    51.3   64.29   87.69   332.1
17 singular values discarded:  -2.26e-15  0  2.52  3.08  4.59  5.12
    7.13  7.64  10.6  12.3  15.4  21.1  23.7  28  29.93  41.04  50.47
```

variable	coeff	variable	coeff		
c0	0.15532	Ile	0.025483		
Ala	0.14642	Pro	-0.024492		
Glu	0.1381	Met	0.024106	Norm of raw data	
Thr	0.11948	His	0.021916	48	
Tyr	0.10543	Trp	0.018428	norm of residuals	
Ser	0.077274	Cys	0.015258	23.593	
Leu	0.049675	Gln	-0.0069745	norm when all SV used	
Lys	0.044517	Phe	-0.0021342	13.451	
Gly	-0.03915	Val	0.00070086		
Asp	0.035248	Arg	0		
Asn	0.027317				

With nine and four eigenvalues, the solution coefficients are smaller and the residual errors are only marginally larger.

Using only the two largest SVs gives small coefficients but a large residual error.

```
Results of SVD analysis  (108 data points, 21 independent variables)
2 singular values used:    87.69   332.1
19 singular values discarded:  -2.26e-15  0  2.52  ...  51.3   64.29
```

variable	coeff	variable	coeff	
Ala	0.15761	Phe	0.026027	
c0	0.14576	Ser	0.020734	
Lys	0.07147	His	0.015757	Norm of raw data
Glu	0.06007	Pro	0.015008	48
Asn	0.05903	Ile	0.014787	norm of residuals
Gly	0.056426	Cys	0.010712	25.506
Leu	0.044367	Met	0.0071074	norm when all SV used
Thr	0.041852	Gln	0.0070112	13.451
Tyr	0.039485	Trp	0.0063071	
Asp	0.03759	Arg	0	
Val	0.03744			

If we compare the top helix formers (with largest coefficients) and the top helix breakers (with lowest coefficients) we have the following results.

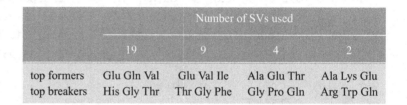

	Number of SVs used			
	19	9	4	2
top formers	Glu Gln Val	Glu Val Ile	Ala Glu Thr	Ala Lys Glu
top breakers	His Gly Thr	Thr Gly Phe	Gly Pro Gln	Arg Trp Gln

Even if we ignore the extreme case of $k = 2$, our analysis did not work too well, since we have a great variability in the solutions depending on the cutoff point. This had to be expected for the very short AA training set used.

For further information try the interactive exercise "SVD – a concrete example."

4 Secondary structure prediction using least squares and best basis

4.1 Definition of "best basis"

In the previous chapter we solved the full least squares problem and retained only the most significant contributions to the solution (by selecting the largest singular values). In this chapter we will choose which independent *variables* (i.e. amino acids) to keep in the solution. This is another way of getting only the most significant contributions to the solution.

Our problem is now to identify the most significant k variables, i.e. k of the n columns of $\mathbf{A}$ which result in a minimal approximation error $\|r\|$. (Remember the size of $\mathbf{A}$ is $m \times n$, with $m \gtrsim 10\,000$ data points and $n = 21$ parameters.) This problem of finding the "best" k variables is called the best basis (BB) problem. This is a really difficult problem, in fact it is NP-complete. Solving it by brute force would require $\mathcal{O}(\binom{n}{k}k^3)$ operations (Table 4.1), since we have to check all possibilities by first choosing k out of n rows then solving the corresponding least squares problem using a gaussian elimination, requiring $\mathcal{O}(k^3)$ operations.

Looking at Table 4.1, we can see that it is marginally possible to use the brute force approach for our problem, since we have $n = 21$ (20 AAs and c_0). For $n > 21$ it quickly becomes impossible. For other problems with large n and k we definitely have to use approximation algorithms.

4.1.1 Stepwise regression

A simple approximation to best basis is stepwise regression, a method used in statistics. One form of stepwise regression, forward selection, starts with an empty basis

n	k	$\binom{n}{k}$	k^3	$\binom{n}{k}k^3$	
10	5	252	125	31 500	a reasonable number
21	10	352 716	1000	352 716 000	borderline
40	20	137 846 528 820	8000	$\approx 1103 \times 10^{12}$	impossible

Table 4.1 Number of operations to find the best k of n columns

and includes one independent variable at a time. The variable added is the one which, of all the remaining ones, reduces the error by the most. When the basis has j variables and we want to increase it to $j + 1$, there are $n - j$ candidates. For each of these candidates we have to solve a least squares problem of size $j + 1$ and find the norm of the error. Then we can select the variable which produces the smallest norm as our $(j + 1)$st variable. These calculations normally require $(n - j) \cdot \mathcal{O}((j + 1)^3)$ operations, i.e. $\mathcal{O}((j + 1)^3 (n - j))$. The total work for a basis of size k is

$$\sum_{j=0}^{k-1} \mathcal{O}((j + 1)^3 (n - j)) = \mathcal{O}(nk^4).$$

A second form of stepwise regression is backward selection, i.e. starting with all variables in the basis and at each step removing the variable which increases the error the least. In this case the total complexity is

$$\sum_{j=k+1}^{n} \mathcal{O}((j - 1)^3 \cdot j) = \mathcal{O}(n^5).$$

This process can be improved by using algorithms which compute solutions to least squares problems from previous solutions which differ in only one basis variable. In this case the compexities become $\mathcal{O}(k^3 n)$ and $\mathcal{O}(n^4 - k^4)$ respectively.

Forward and backward stepwise regression are not optimal algorithms, they are what is normally called *greedy algorithms*. They choose the optimal variable for each individual step, which does not guarantee global optimality. As it turns out, stepwise regression seldom gives the global optimum, the optimizations shown in Section 4.3 give better results.

4.1.2 The difficulty of the problem

The best basis problem is inherently difficult, since it does not allow a "continuous approximation." This is illustrated in the following counterexample.

For a matrix $\mathbf{A}$ with $\dim(\mathbf{A}) = m \times 2k$ set the first k columns $\mathbf{A}_1, \ldots, \mathbf{A}_k$ to random values. Set b to be the sum of these columns. Hence a perfect solution with $\|r\| = 0$ is given by choosing these first k columns. Now choose the next $k - 1$ columns

$A_{k+1}, \ldots, A_{2k-1}$ to be random and choose a random vector ε of very small values (to be used as errors). As a last step, set the final column of A to be $b - \sum_{i=k+1}^{2k-1} A_i + \varepsilon$. Choosing columns $k + 1$ to $2k$ as a solution will give $\|r\| = \|\varepsilon\|$ as error, which could be made arbitrarily small, since we are choosing ε.

So we have two *completely* different sets of solution vectors, one with error zero, the other with an arbitrarily small error of $\|\varepsilon\|$. All other solutions will have very large $\|r\|$, since they will involve at least one missing random column.

Conclusion There is no guarantee that a selection of columns with a small error will be "close" to the optimum.

Although this counterexample shows that close to optimal solutions may not be "neighbors," in real problems it is still useful to explore the neighborhood of a good solution hoping to find a better one.

4.2 Mathematical formulation using the best basis approach

Let $c = \{c_1, c_2, \ldots, c_k\}$ be a set of column indices, i.e. a subset of $\{1, 2, \ldots, n\}$. Since the order of the c_i is not relevant, we will use the natural order $c_i < c_{i+1}$.

Let $F(\{c_1, c_2, \ldots, c_k\}) = \|r\|$ be the residual error of the least squares approximation, when using the k columns $c_1, c_2, \ldots, c_k$. The best basis problem is to find the subset $c \subset \{1, 2, \ldots, n\}$, such that $F(c)$ is minimized.

Let $d \subset \{1, 2, \ldots, n\}$ be a neighbor of c. Mathematically written:

$$d = \{d_1, d_2, \ldots, d_k\} \in N(c) \quad \Leftrightarrow \quad |d \setminus c| = |c \setminus d| = 1,$$

i.e. the sets c and d each contain exactly one element, which is not an element of the other set. Since c and d are sets of column indices, this means that the corresponding matrices will each contain only one column, which is not contained in the other matrix. For example for $k = 4$, $n = 10$ and $c = \{2, 7, 8, 9\}$, $d = \{2, 5, 7, 9\}$ is a neighbor of c.

How many neighbors $N(c)$ does a set c have? The answer is $|N(c)| = k(n - k)$, since each of the k columns can be replaced by one of the $n - k$ not used columns.

For example:

$$n = 4, k = 2, c = \{1, 3\}$$
$$N(c) = \{\{1, 2\}, \{1, 4\}, \{2, 3\}, \{3, 4\}\} \quad \text{and} \quad |N(c)| = 2 \cdot 2 = 4.$$

Our optimization algorithm consists of finding a local optimum using the $N(c)$ neighbor relation. A local optimum will be defined by a point c where the function at all neighbors is not smaller, i.e. $\forall d \in N(c) \implies F(c) \leq F(d)$.

Although we define this optimization for the problem of best basis, the techniques of optimization based on neighbors are quite general and are often very successful.

4.3 Algorithms for finding local minima

We will find local optima by starting at a random point c and going through its neighbors as long as the function decreases. Two general strategies are possible: early abort (EA) and steepest descent (SD).

4.3.1 Early abort

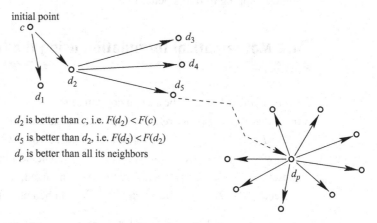

d_2 is better than c, i.e. $F(d_2) < F(c)$
d_5 is better than d_2, i.e. $F(d_5) < F(d_2)$
d_p is better than all its neighbors

Figure 4.1   Schematic diagram of the early abort algorithm.

This strategy searches through all the neighbors and restarts at the first new point it finds with a lower value (see Figure 4.1).

Algorithm

```
        c := random set of column indices;
        t := F(c);
restart: for d in N(c) do
            t1 := F(d);
            if t1 < t then
                c := d;
                t := t1;
                goto restart;
            fi;
        od;          # ⇒ c is a local optimum.
```

4.3.2 (Discrete) steepest descent

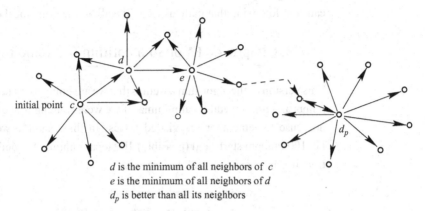

d is the minimum of all neighbors of c
e is the minimum of all neighbors of d
d_p is better than all its neighbors

Figure 4.2 Schematic diagram of the steepest descent algorithm.

This strategy explores all the neighbors of a point c and chooses the one with the minimal value as the next point, provided it is better than c (see Figure 4.2).

Algorithm

```
c := random set of column indices;
t := F(c);
change = true;
while change do
     change := false;
     for d in N(c) do
          t1 := F(d);
          if t1 < t then
               c := d;
               t := t1;
               change := true;
          fi;
     od;
od;                 # ⇒ c is a local optimum.
```

In both methods, the most expensive step is the computation of $F(c)$, which requires solving a LS problem. An improvement can be made by avoiding computations of points already inspected.

Definition A *path* is a sequence of neighbors, which ends at a local optimum. Paths are directed.

It is important to keep track of paths and their visited neighbors because if we search again and are on a new path and we move onto an already visited point, we can stop following that path, as we will walk to the same local optimum.

4.3.3 Improved EA or SD algorithm (avoiding repeated paths)

This version of the algorithm assumes that we are going to try several random starting points and find several local minima. If we were running it only once, it would make no sense to remember the visited nodes. In this example we run 100 iterations, i.e. 100 random starts and (possibly) 100 local minima. We define visited[c] as true, if c has been in a path before.

```
            for iter from 1 to 100 do
                c := random set of variables;
                if visited[c] then next fi;
                visited[c] := true;
                t := F(c);
restart:        for c1 in N(c) do
                    t1 := F(c1);
                    if t1 < t then
                        c := c1;
                        t := t1;
                        if visited[c] then break fi;
                        goto restart;
                    fi;
                    visited[c1] := true;
                od;
                LocalMinim[iter] := c;
            od;
```

Remark next is a Maple command which skips to the next iteration, equivalent to continue in C. break will terminate the do-loop.

The algorithm shown is EA, it is a simple transformation to use SD.

4.3.4 Exploring the solution space

Both EA and SD find local optima of the problem. Because of the complexity of the problem and the limited number of neighbors, we cannot give any guarantees that the local optimum is a global optimum. Normally it is not. EA and SD become much more effective when we select several random starting points and find their local optima (as above in the improved version of EA). Not only may we get a better local optimum, but we may also get an idea of the complexity of the solution space.

For example, suppose we run the optimization starting at 100 different random points.

Scenario A: we get 95 times one optimum and 5 other times another one, slightly worse. This is a very good case as the solution space seems to have very few local optima.

Scenario B: we get 100 different optima. This is a very bad case, clearly the space is full of "depressions" like sand dunes in the Sahara. Finding the global minimum is just a matter of good luck.

Because of the above we will normally run the optimization several times starting at different random points. For EA it also makes sense to go through the neighbors in a random order, rather than in a prescribed, fixed order. See Section 8.8.2 for more details of this exploration.

Starting at a random point and descending to a local optimum allows us to run this algorithm in parallel. Each processor will run its own optimization path. This is a very simple and effective form of parallelism, sometimes called "embarrassingly parallel." Since optimization problems may be very costly in terms of computation time, being able to optimize in parallel is a big advantage.

With EA, few neighbors are visited in the first steps. The last step has to visit all neighbors, i.e. it is expensive. All the steps of SD are as expensive as the last step. Normally we expect to do fewer steps in SD, but the computation saving in the early steps normally favors EA.

In both EA and SD, after the first step, we should avoid exploring the neighbor from which we came, which is certainly worse. SD will share $n - k$ neighbors with its predecessor, where n is the total number of variables, of which we select k. (Prove it as an exercise!)

4.4 The neighbor function

A good neighbor function has to satisfy some properties to be usable and efficient.

Reachability There should be a path (a sequence of neighbors) that joins any two arbitrary points in our space. This path should be of polynomial (preferably linear) length in the size of our problem. In our case any two solutions are linked by a path of neighbors of length less than or equal to k. An implication of the above is that the neighbor function connects all solutions.

Number of neighbors The number of neighbors of a given solution point should also be polynomial in the size of the problem (linear or quadratic). If it is too large, then the search for a better neighbor will take too long. On the other hand too few neighbors will produce optimization paths that end in simpler local optima. In our case, the number is $k(n - k)$ which is quadratic in the size of our problem.

Smoothness of neighbors It is expected, although difficult to guarantee, that neighbors are of similar quality. That is to say that from a good solution we will find similarly good (perhaps better) solutions among its neighbors. If this cannot be guaranteed, the optimization procedure is equivalent to a random search.

4.5 The confidence interval

Within each basis it is easy to order the variables according to their contribution. For each variable we compute

$$t_i = F(c \setminus \{c_i\})^2 - F(c)^2,$$

where c_i is the ith element in the solution vector c. t_i represents the norm increase when column c_i is not used. Intuitively, the larger the norm increase, the larger the importance of c_i. If c_i is important and we remove it, the error of the solution will increase significantly.

This suggests that t_i is a good criterion to order the independent variables in the solution, most important first, least important last. This ordering is useful in the following ways.

Per se In our example we would want to know which is the most significant amino acid influencing helix formation.

Relative contribution This can help us to decide when to reduce the number of independent variables, e.g. if the least significant variable reduces the error only by 0.1% of the total error, it is unlikely to be relevant.

Relative to random In Section 4.6 we use the idea of adding random columns as independent variables. These random columns/variables serve as witnesses of the limit of significance. Hence the ranking of a random column/variable marks the limit of significance.

The following output shows our helix prediction problem for $k = 6$ ordered by t_i (last column) and with the variance of the computed coefficients (see the following subsection).

```
Results of SVD analysis
21 independent variables, 6 used here
Norms of:      raw data      residuals
                   48           15.533
6 singular values used: 7.345 14.03 32.09 38.98 60.55 223
          variable        coeff/stdev      norm decrease
            c0          0.4723 +- 0.0774       5.67
            Glu         0.3224 +- 0.0710       3.139
            Gly        -0.2929 +- 0.0733       2.431
            Thr        -0.2594 +- 0.0729       1.929
            His        -0.4360 +- 0.1388       1.503
            Ala         0.1194 +- 0.0496       0.8829
```

4.5.1 Analysis of the variance of the coefficients

Suppose the correct solution to our problem is a vector s so that $\mathbf{A}s = \hat{b}$. Now, due to random errors in the data, we do not get $\hat{b}$, but $b = \hat{b} + \varepsilon = \mathbf{A}s + \varepsilon$, where ε is a vector of dimension m of errors of the dependent variable. The solution we obtain is x (if we ignore the SVD analysis and solve the normal equations), where

$$\mathbf{A}^T\mathbf{A}x = \mathbf{A}^T b$$

or $\quad\quad x = (\mathbf{A}^T\mathbf{A})^{-1}\mathbf{A}^T b = \mathbf{W}b,$

with $\quad\quad \mathbf{W} = (\mathbf{A}^T\mathbf{A})^{-1}\mathbf{A}^T$ a matrix of dimension $n \times m$,

$$x = \mathbf{W}(\mathbf{A}s + \varepsilon) = s + \mathbf{W}\varepsilon$$

or $\quad\quad x - s = \mathbf{W}\varepsilon.$

This means that the errors in estimating the coefficients of the real solution $x - s$ are linearly dependent on the errors in the data. This holds under the assumption that the correct model is indeed a linear model. The implications of this linear dependence are significant, for example if we cut the errors in half, the errors in the coefficients will also be cut in half.

It is a common assumption that the errors ε_i are independent of each other, unbiased, i.e. $E[\varepsilon_i] = 0$, and normally distributed with identical variances, i.e. $\varepsilon_i \in N(0, \sigma^2)$. Under these conditions, the error for the ith coefficient is

$$x_i - s_i \in N(0, (\mathbf{A}^T\mathbf{A})_{ii}^{-1}\sigma^2).$$

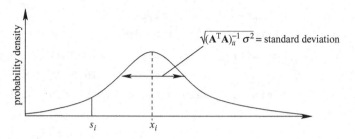

Figure 4.3   **Normal distribution of errors.**

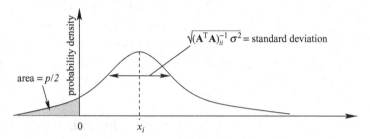

Figure 4.4   **The probability that x_i is not part of the solution is $\Pr(x_i \leq 0) = p/2$.**

In other words, the main diagonal of the matrix $(\mathbf{A}^\mathrm{T}\mathbf{A})^{-1}$ gives the factors that will multiply the variance of the errors in $\boldsymbol{b}$.

Once we know that the computed coefficients have a normal distribution (Figure 4.3) we can conclude several confidence measures.

(i) We can use some standard confidence limits, e.g. 95% and bound the range of values for s_i:

$$|x_i - s_i| < 1.96\sqrt{(\mathbf{A}^\mathrm{T}\mathbf{A})^{-1}_{ii}\sigma^2}.$$

(ii) We can use the standard deviation $\sqrt{(\mathbf{A}^\mathrm{T}\mathbf{A})^{-1}_{ii}\sigma^2}$ or the relative standard deviation $\sqrt{(\mathbf{A}^\mathrm{T}\mathbf{A})^{-1}_{ii}\sigma^2}/x_i$ as a measure of quality of the fit (in this case, the smaller the better) and order the independent variables according to this value.

(iii) We can estimate whether this variable could be zero (i.e. not be part of the solution, Figure 4.4) by computing the probability p that $|x_i - s_i| \geq |x_i|$.

4.6 Confidence analysis with random columns

The best basis approach allows us to do a different type of confidence analysis which is not possible with the classical SVD based LS. This confidence analysis is also very effective and powerful.

Suppose that we have our data matrix $\mathbf{A}$ of size $m \times n$ as before, from which we want to select the most significant $k < n$ columns. We will add n addditional columns with random values, resulting in a matrix $\mathbf{A}^*$ of size $m \times 2n$.

We now select the best k columns with the BB algorithm. If any of the new (random) columns shows up as part of the solution, this is a clear signal that we may have a problem, in particular that k may be too large.

To understand the reasoning behind this idea we will give an intuitive argument. Assume that all the data are random, the initial n columns as well as the additional ones. Then the highest ranking column will be from the first group or second group with equal probability. In other words, 50% of the time the highest ranked random column from the second group will be at the top. It is easy to conclude that if we have a mixture of non-random and random columns, then some portion of the time, $(n + k^*)/2n$, the top ranking random column (second group) will be at the top of the ranking of random columns and hence mark the limit of useful/random variables. The k^* in the above formula is the number of columns/variables which are significant (non-random).

For all practical purposes, when we do BB with additional random columns, the highest ranking random column marks the limit of the significant variables. And if this is ranked as the first variable, then nothing is significant! The number k should be adjusted accordingly, so that $k + 1$ is the highest ranking random column half of the time.

4.7 Final observations on the least squares methods

For the LS methods we obtain a linear equation which should predict the existence of a helix. In this section we describe how to refine this predictor function and how to validate the whole process.

4.7.1 Refining the predictor function

We still have a minor technical problem to resolve. We want a decision function or a function which gives us a one if there is a helix and a zero if not:

$$f(\ldots) = \begin{cases} 1 & \text{if helix} \\ 0 & \text{if non-helix.} \end{cases}$$

But the result of our linear function may be any value, not even necessarily between zero and one. So we have to define a "barrier" d_0 which gives us a decision for helix/non-helix.

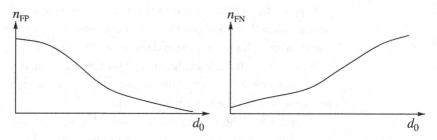

Figure 4.5   Number of *false positive* n_{FP} and number of *false negative* n_{FN} depending on d_0.

Example: $f(2E + 2I + Q) \approx 1.24$

$f(\ldots) \geq d_0 \Rightarrow$ helix

$f(\ldots) < d_0 \Rightarrow$ non-helix.

How should we determine d_0? We could set $d_0 = 1/2$, this would be equal to rounding the value of $f(\ldots)$ and is reasonable but not necessarily optimal. A better approach is to count n_{FP}, the number of *false positives*, and n_{FN}, the number of *false negatives*, for each possible d_0 and select the best one (Figure 4.5). We have a false positive if our function $f(\ldots) \geq d_0$ predicts that there is a helix, but in reality **there is no** helix. And a false negative occurs if $f(\ldots) < d_0$ predicts no helix, but in reality **there is** a helix.

We can now choose a d_0 so that the weighted sum of n_{FP} and n_{FN} is minimal. To do this efficiently we sort all values of $f(\ldots)$. First we set d_0 to the largest value of f, which is equal to $n_{FP} = 0$ and $n_{FN} = \#$(helices), calculate n_{FP} and n_{FN}, multiply them with their weighting factors w_{FP} and w_{FN} respectively and add them together (see Section 3.3.1 for the need for weighting):

$$\text{weighted number of errors} = w_{FP} \cdot n_{FP} + w_{FN} \cdot w_{FN}.$$

Now we set d_0 to the next lowest value of $f(\ldots)$, and continue this process for all the sorted $f(\ldots)$ in descending order down to the smallest. Keeping track of the smallest weighted error we eventually find d_0. Notice that the number of false positives or negatives can only change when d_0 is equal to some $f(\ldots)$ value. So we only analyze these transition values for d_0.

4.7.2 Validation

We use part (half) of the data to fit the parameters and the second part to validate the model. It is very tempting (and absolutely incorrect!) to conclude the quality of the model from the fitted data. In particular it is recommended that the validating data be used only once, i.e. do as much work as you want with the fitting, but once a model is computed, use the validation data to test it. These validation data should not be used again for validation purposes. The reasoning behind this rather radical approach is that if we use our validating data many times (to reject various models), de facto we are using them as training data.

Testing the model against the validation data will tell how accurate our predictions are. Remember that 50% (weighted) accuracy is achieved by random choices, so that is no accuracy at all.

The acceptable accuracy depends on the problem. For the SSP anything above 75% is good. Other applications may require 95% while others might be happy with 51%.

The crucial test is to insure that our model is predicting above a certain level p. E.g. $p = 0,75$ for SSP. Assuming that we have N independent test in our validation set and that N_C are predicted correctly (and $N - N_C$ incorrectly), then if

$$\frac{N_C - N_p}{\sigma(N_C)} > 1.96 \quad \text{or}$$

$$N_C - N_P > 1.96\sqrt{N_P(1 - p)}.$$

We can conclude with 95% confidence that our predictions are correct and at a level of p. See Appendix A1.5.

4.7.3 Other models

Up to now we have looked at functions that used a window like

$$f(a_{-2}, a_{-1}, a_0, a_1, a_2) = a_{-2} + a_{-1} + a_0 + a_1 + a_2 \quad \text{(symmetric)}$$
$$f(a_0, a_1, a_2) = a_0 + a_1 + a_2 \quad \text{(asymmetric)}.$$

The window could be longer or shorter.

Another idea is to use *weighted functions* like

$$f(a_{-2}, a_{-1}, a_0, a_1, a_2) = w_{-2}a_{-2} + w_{-1}a_{-1} + w_0a_0 + w_1a_1 + w_2a_2$$
$$= \sum_{i=-2}^{2} w_i a_i.$$

The problem of this approach is that if we want to optimize the weights too, it leads to a non-linear least squares minimization, which is much more difficult to solve.

Still another approach is to use some properties of the amino acids which have numerical values instead of just counting amino acids. Possibilities are

- hydrophobicity index $H(\text{AA})$,
- molecular weight/size $W(\text{AA})$,

- charge $C(AA)$,

$$f(a_{-2}, a_{-1}, a_0, a_1, a_2) = h_{-2}H(a_{-2}) + w_{-2}W(a_{-2}) + c_{-2}C(a_{-2})$$
$$\cdots$$
$$+ h_2 H(a_2) + w_2 W(a_2) + c_2 C(a_2).$$

In this case we would have 15 h, w and c as unknowns (for three properties and window size 5) and a linear least squares problem.

4.8 Linear classification/discrimination

The previous sections showed methods to construct a linear function which approximates helices (1) and non-helices (0) numerically. In matrix notation, for the data matrix A we compute a vector x and a value d_0 with $0 \le d_0 \le 1$ such that $A_i x > d_0$ is an α-helix, and $A_i x \le d_0$ is not an α-helix, where A_i is the ith row of A and corresponds to the ith observation.

The problem can be treated in a more direct way, and instead of approximating how close the values of $A_i x$ are to 0 or 1, we could minimize the total number of errors made in the prediction. That is, given A and b find a vector x and a value d_0 so that the sum of the number of times that $A_i x > d_0$ when $b_i = 0$ plus the number of times that $A_i x \le 0$ when $b_i = 1$ is minimized.

The cases when $A_i x > d_0$ and $b_i = 0$ are called false positives and the cases when $A_i x \le d_0$ when $b_i = 1$ are called false negatives. In these terms we want to find x and d_0 which minimize the (weighted) sum of false positives and false negatives. It is often the case that we want to weight the positives and the negatives, if their numbers are very different (e.g. 10 positives versus 1000 negatives). This is discussed in Section 4.7.1.

The problem as presented is very difficult, it is NP-hard even for integer matrices A. So we do not expect to solve it exactly, we will be happy with an approximation (which is what we were doing with SVD and Best Basis).

In statistics this problem is resolved in a particular way and it is called *Fisher's linear discriminant*.[1] The view used for Fisher's linear discriminant is to maximize the distance between the averages of the positives and of the negatives relative to the variance of the whole sample. This is a desirable property and geometrically meaningful, but does not guarantee that we achieve our objective of minimizing the

---

[1] Wikipedia: http://en.wikipedia.org/wiki/Linear_discriminant_analysis
Interactive applet: http://www.ml.inf.ethz.ch/education/lectures_and_seminars/annex_estat/Classifier/JFishersLinearDiscriminantApplet.html

number of incorrect predictions. The Fisher discriminant reduces to the solution of an eigenvalue/eigenvector problem.

4.8.1 Trying to do better

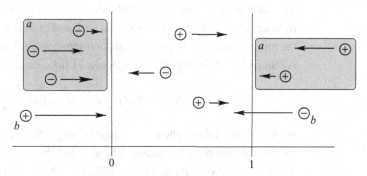

Figure 4.6 Trying to improve the approximation.

The main problem with the least squares methods is that they try to approximate *all* positives to 1 and *all* negatives to 0. This is in principle a good goal but fails in two cases:

 (i) a point which is correctly classified but whose value is too large (already larger than one or smaller than zero – cases labelled "*a*" in Figure 4.6), where any change is useless, or
(ii) a point which is incorrectly classified and whose value is hopelessly far from d_0 (like both points marked "*b*" in Figure 4.6).

The arrows are shown to indicate what LS tries to achieve, i.e. move the points in the right direction (the longer the arrow, the larger the "pressure" to move the point). Recall that we are minimizing the square of the errors, so the "pressure" to move in the correct direction is proportional to the distance to the correct point, zero or one. It is clear that it is useless to move the points marked "*a*". It is also clear that it is hopeless to move the points marked "*b*", they are too far away.

An excellent strategy to improve a classification (obtained in any way) is to identify these outliers (the *a* and the *b*) and redo the classification without them. This gives an algorithm for improvement or refinement which has the following steps.

Algorithm Refining a linear classification: $(\mathbf{x}, d_0) :=$ initial classification.

1. Select all the points for which $|\mathbf{A}\mathbf{x} - d_0| < s$ (all points close to the decision point, not outliers).

2. For these points compute a direction vector d, such that $\|Ad - (2b - 1)\|$ is minimal. (Approximating to $2b - 1$ is equivalent to approximating the negatives to -1 (move left) and the positives to $+1$ (move right).)

3. Explore $x + hd$ for small positive h to see whether the weighted number of false negatives and positives decreases. If so, replace $x \leftarrow x + hd$ and restart at step 1.

The main idea of step 2 is to find a direction which will move the positives to the right and the negatives to the left. This will tend to produce a minor rearrangement of the points around d_0, hopefully with an improvement. The exploration of the h-space is relatively simple. By solving each row of the data for the corresponding h, i.e.

$$A_i(x + h_i d) = d_0 \quad \text{or} \quad h_i = \frac{d_0 - A_i x}{A_i d}$$

we find all the values where a data point changes from being correct to incorrect. We can sort all the positive values of h_i and find the optimal h with a sequential scan of these transition values. The algorithm can be tried with many values of s or by choosing s so that a particular fraction of the data points is selected. Other variations on how to select the candidate points are also possible.

This is an application of the constructor–heuristic–evaluator (CHE) approach, discussed in more detail in Section 8.2.2. The constructor could be the LS approximation to 0/1, the evaluator is simply the weighted sum of false negatives and positives which should be minimized. The method above (and other similar possible improvements) is the heuristics.

4.8.2 Constructing initial solutions using the sigmoid function

The sigmoid function (Figure 4.7) is defined by

$$s(x) := \frac{1}{1 + e^{-x}}.$$

Sigmoid means that the shape is similar to an "s," a stretched "s."

We can see that the limit of the sigmoid function for $x \to \infty$ is $\lim_{x \to \infty} s(x) = 1$ and $\lim_{x \to -\infty} s(x) = 0$. Different constants multiplying the argument make the transition sharper or smoother. Furthermore, for arguments far from zero, the sigmoid function does not change much, it becomes very flat (approximately one for positive arguments, approximately zero for negative arguments).

These properties are very suitable for transforming our decision problem into a continuous numerical problem. This is done by minimizing the function

$$\sum w_i(s(A_i x - d_0) - b_i)^2,$$

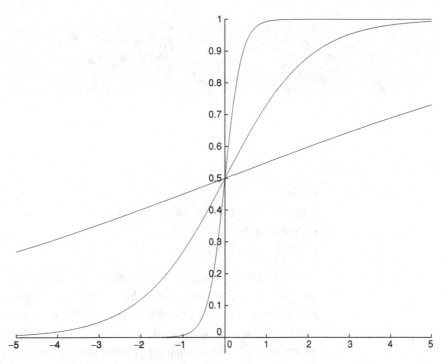

Figure 4.7   **The sigmoid functions $s(5x)$, $s(x)$ and $s(x/5)$.**

i.e. finding (x, d_0) that will give a minimal value to the sum of squares of the (decision) errors. The weights w_i are used as needed to balance the influence of the positives and negatives.

The problem is no longer linear, so numerical minimization is needed. The minimization should be started from the point $x = 0, d_0 = 0$. The methods in Appendix A1.3 can be used to solve this problem numerically. The sigmoid method, as well as LS and Fisher's discriminant produce good solutions but not necessarily optimal ones. All of them can be improved with the heuristic mentioned above.

## 5

Secondary structure prediction with learning methods (nearest neighbors)

Topics

- Nearest neighbor searching
- Clustering
- Binary search trees

Learning methods are, in general, methods which adapt their parameters using historical data. Nearest neighbors (NN) is an extreme in this direction, as all training data are stored and the most suitable part is used for predictions. Unless there are contradictory data, every time that NN sees the same information as in the training set, it will return the same answer as in the training set. In this sense, it is a perfect method, as it repeats the training data exactly.

Generalities of nearest neighbor methods (Figure 5.1)

(i) NN methods extract data from close neighbors, which means that a distance function has to be defined between the data points (to determine what is close and what is distant). The data columns may be in very different units, for example kg, years, mm, US$, etc., so the data have to be normalized before computing distances.

(ii) Finding neighbors for small sets of training data (e.g. $m < 1000$) is best done with a sequential search of all the data. So our interest here is in problems where m is very large and we have to compute neighbors many times.

(iii) As opposed to best basis where extra variables were not obviously harmful (in some cases random columns were even useful), NN deteriorates if we use data columns which are not related to our data.

(iv) NN methods have a close relationship with clustering for which our main algorithm for NN can also be used.

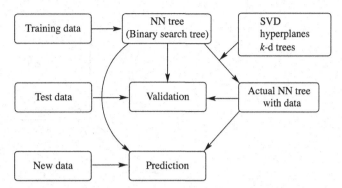

Figure 5.1 Scheme for modelling with nearest neighbors.

5.1 Formulation of the problem

5.1.1 The general idea

The main idea of this method is to store all data points of known helices and non-helices. For a new data point we search in this database for the closest data points, called "nearest neighbors," and we decide according to the values of these neighbors.

General algorithm

(i) Describe each data point $c_1, \ldots, c_m$ by a mapping of the n-dimensional vector of numerical values (coordinates) to the value(s) to predict:

$$
\begin{aligned}
c_1 &= (c_{11}, c_{12}, \ldots, c_{1n}) &\rightarrow& \quad v_1 \\
&\vdots& &\quad \vdots \\
c_m &= (c_{m1}, c_{m2}, \ldots, c_{mn}) &\rightarrow& \quad v_m.
\end{aligned}
$$

This mapping is viewed as a function in n-dimensional space (Figure 5.2, next page), which is only defined for the points for which we have data.

(ii) We define a distance function between points c_i and c_j by:

$$
d(c_i, c_j) = \sqrt{\sum_{k=1}^{n} w_k (c_{ik} - c_{jk})^2} \quad \text{with} \quad w_k > 0 \quad \text{for } k = 1, \ldots, n.
$$

The weights w_k normalize the distances. The goal of this normalization is to have points equally distributed in all dimensions.

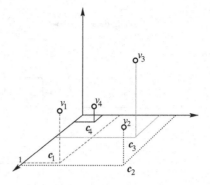

　Data points c_i and their values v_i for $n = 2$.

(iii) We store every point of the training set in an appropriate data structure (NN tree).

(iv) To predict the value of a new point we find its nearest neighbors and interpolate the result from their values.

5.1.2 Application to helix prediction

The first question we have to answer is the dimension of the problem. The dimension n depends on the window size we use. If we use hydrophobicity, molecular weight and charge per amino acid as described in the previous chapters,

$$\overbrace{(H(A), S(A), C(A)}^{\text{first amino acid}}, \ldots, \overbrace{H(V), S(V), C(V))}^{\text{last amino acid}} \underbrace{}_{n\text{-dimensional data point}} \Rightarrow \text{yes/no},$$

then we have a window of five amino acids and three values per AA, i.e. we have $n = 15$.

The distance is defined as above in (ii). We will use the weights to normalize for the different units, for example the hydrophobicity is usually between -1 and $+1$ while the molecular weight is in daltons and between 50 and 150. Clearly we cannot combine them directly ($w_i = 1$) as this would mean that one measure will dominate and others will have no influence. Weights that normalize the different dimensions to insure that all have the same variance are given by

$$w_i = \frac{1}{\sigma_{x_i}^2} \quad \text{(where the } x_i \text{ indicates the } i\text{th dimension)}$$

or by the Z-transform (see Section 5.5.2).

5.2 Searching and data structures for NN problems

The nearest neighbor problem can be solved in several ways. We will look at the following cases/algorithms.

(i) For relatively few training points, or for very few predictions, a sequential search solves the problem exactly.

(ii) For very few dimensions ($n \leq 3$), quad trees provide an elegant and efficient solution.

(iii) For the difficult cases with high dimension and large number of training points, we shall develop a variant of k-dimensional search trees.

5.2.1 Sequential search

We store every data point in an array and compute the distance of each point to our data. Then we select the point with the shortest distance.

Complexity: (# points in training set) · (# points in test set) $\approx \mathcal{O}(m^2)$ if our testing data are of the same order of magnitude as our training data. This technique is out of the question for large amounts of data as in the case of helix prediction.

5.2.2 Range searching

Given a set of points in the plane, it is natural to ask which of those points fall within some specified area. "List all cities within 50 miles of Zürich" is a question of this type which could reasonably be asked if a set of points corresponding to the cities of Switzerland were available. This is a NN problem. When the geometric shape is restricted to be a rectangle, the issue readily extends to non-geometric problems. For example, "list all those people aged between 21 and 25 with income between $60 000 and $100 000" asks which "points" from a file of data on people fall within a certain rectangle in the age–income plane. This is a range searching problem.

Extension to more than two dimensions is immediate. If we want to list the ten closest stars to the sun or any other arbitrary star, we have a three-dimensional NN problem, and if we want the rich young people to be tall and female as well, we have a four-dimensional range searching problem. In fact, the dimension of such problems can get very high.

In general, we assume that we have a set of records with certain attributes that take values from some ordered sets. Finding all records in a database that satisfy specified range restrictions on a specified set of attributes is called *range searching*, and is a difficult and important problem in practical applications. Although similar, range searching algorithms are not directly applicable to NNSearch.

5.2.3 Quad trees

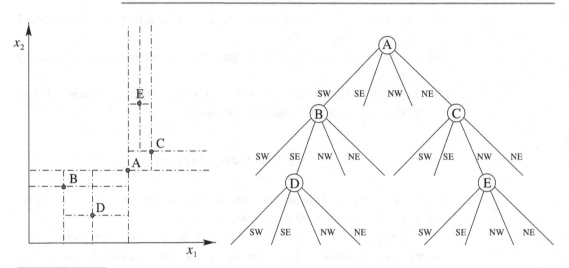

Figure 5.3 Example of a quad tree data structure.

Quad trees in n dimensions split the searching space in 2^n subspaces at each node. That is, for each internal node, the point stored at the node splits the remaining data points according to all possible subspaces. A quad tree for $n = 1$ is identical to a binary search tree.

A quad tree for $n = 2$ splits the two-dimensional space in four subspaces at each internal node – therefore the name quad tree. The four descendants are usually called north-east (NE), north-west (NW), south-east (SE) and south-west (SW). The data point in the node is exactly in the middle of this partition, see Figure 5.3.

Quad trees are not practical for our helix prediction problem: the quad trees for n dimensions have internal nodes with 2^n descendants. Hence these trees are totally impractical for large n. Quad trees are effective for $n \leq 3$ only.

5.2.4 k-d Trees (k-dimensional trees)

The generalization from binary search trees to k-d trees is quite straightforward: simply *cycle through the dimensions* while going down the tree, i.e. for the root use the first dimension for the decision about the subtrees, for the second level use the

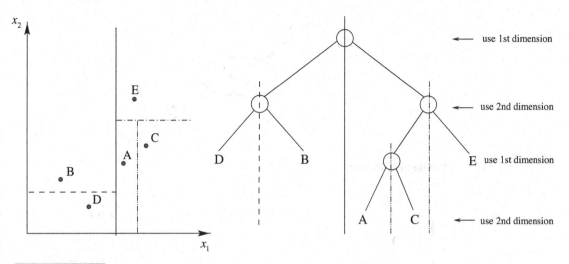

Figure 5.4 An example of a k-d tree with $k = 2$: alternating vertical splits (left is left point) and horizontal splits (left is lower point).

second dimension etc., see Figure 5.4. With random data, the resulting trees have the same good characteristics as binary search trees.

For very high dimensions, say 20–40, the k-d algorithm has to be modified to be useful. If we use it as defined, cycling over all the dimensions, we will normally exhaust the tree before we even have used some of the dimensions. Hence the tree will be insensitive to the dimensions not used. Another problem arises if the most important dimension is for example the seventeenth. Then we will split 16 times (2^{16} subtrees) before we use our data. Clearly k-d trees were not designed for a very high number of dimensions. The solution to this problem is based upon using a linear combination of dimensions at each level (instead of using a single one). The modified data structure is called a nearest neighbor tree.

5.2.5 Nearest neighbor trees (NN trees)

At every node in a NN tree we will store a vector $\boldsymbol{\alpha}$ (the unit normal vector of a hyperplane) and a threshold value α_0 (where α_0 is the distance of the chosen hyperplane from the origin). The Hessian normal form of hyperplanes is $\boldsymbol{\alpha x} = \alpha_0$. This hyperplane will be used to split the descendants into two subtrees (Figure 5.5, next page).

When we are inserting or searching a point with coordinates x and we arrive at this node, we compute the inner product $\boldsymbol{\alpha} \cdot \boldsymbol{x}$ and go left or right depending on whether

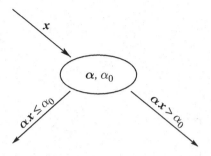

Figure 5.5   A node of a NN tree.

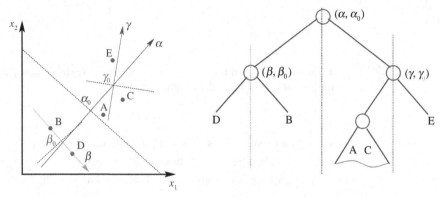

Figure 5.6   Our $2d$-example with hyperplanes (dotted lines), normal vectors α, β, γ and the resulting data structure. How would the normal vector δ, splitting A and C have to be chosen?

$\alpha x \leq \alpha_0$ or $\alpha x > \alpha_0$

$$\boldsymbol{\alpha} \cdot \boldsymbol{x} = \alpha_1 x_1 + \alpha_2 x_2 + \cdots + \alpha_n x_n \begin{cases} \leq \alpha_0 & \implies \text{go left} \\ > \alpha_0 & \implies \text{go right.} \end{cases}$$

Figure 5.6 explains the NN tree.

A hyperplane in n dimensions is defined by an equation $\boldsymbol{\alpha} \cdot \boldsymbol{x} = \alpha_1 x_1 + \alpha_2 x_2 + \cdots + \alpha_n x_n = \alpha_0$.

What is a good α_0? We can select α_0 so that we get a balanced tree. So a good α_0 will be one which splits the nodes $\boldsymbol{x}$ into subtrees of the same (or nearly the same) size (Figure 5.7).

What is a good hyperplane? A good hyperplane is one for which the $\boldsymbol{\alpha}\boldsymbol{x}$ are spread apart as much as possible (Figure 5.8). This has a formal definition: we compute $\boldsymbol{\alpha}$, so that $\boldsymbol{\alpha}\boldsymbol{x}$ (for the set of coordinates $\boldsymbol{x}$ in our subtree) has largest variance.

This is a relatively well known problem in statistics. The largest variance is obtained with an $\boldsymbol{\alpha}$ which is the eigenvector with the largest eigenvalue of the covariance matrix

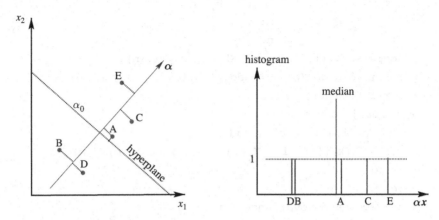

Figure 5.7   A good choice of hyperplane. (Remember that α is the normal vector of the hyperplane, i.e. perpendicular to the plane.)

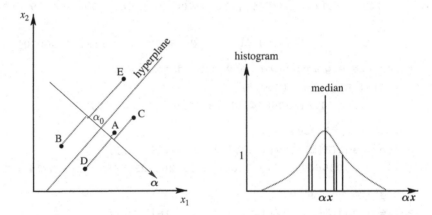

Figure 5.8   A bad choice of hyperplane. Note that all the points cluster their projections onto α together.

of x. The covariance matrix is defined by

$$\mathrm{Cov} = \left(c_{ij}\right), \quad \text{where} \quad c_{ij} = E\left[(x_i - \bar{x}_i)(x_j - \bar{x}_j)\right]$$

with E the expectation value.

Intuitively, using the α that maximizes the variance of αx is like using the direction which is best aligned with the bulk of the points (α in Figure 5.6).

Example of a Maple session

```
> X:= [[1,11,25],[12,0,15],[10,2,19],[17,5,6],[2,3,18]]:
> m:= nops(X); n:= nops(X[1]);
```

```
                       m:=5
                       n:=3
> Cov:= array(1..n,1..n): Xb:= array(1..n):
> for i to n do Xb[i]:= add(X[j,i],j=1..m) / m od:
> for i to n do
>    for j to n do
>      Cov[i,j]:= evalf( add(
>        (X[k,i]-Xb[i]*(X[k,j]-Xb[j])), k=1..m) / m)
>    od;
> od;
# averages and covariance matrix
> print(Xb,Cov);
```

$$
[42/5,\ 21/5,\ 83/5],\ \begin{bmatrix} 37.04000000 & -10.88000000 & -32.84000000 \\ -10.88000000 & 14.16000000 & 9.680000000 \\ -32.84000000 & 9.680000000 & 38.64000000 \end{bmatrix}
$$

```
# compute eigenvalue/eigenvector decomposition:
> evalf( Eigenvals(Cov,V));
          [4.818906844, 10.81601554, 74.20507763]

# lowest variance:
> [seq( add(X[i,j]*V[j,1],j=1..n), i=1..m)];
   [-19.10356001, -18.70605688, -20.22366497, -17.22143672, -13.87966192]

# highest variance:
> [seq( add(X[i,j]*V[j,n],j=1..n), i=1..m)];
   [19.23462308, 2.221176762, 6.826857985, -6.247865674, 11.81962487]
```

5.3 Building NN trees

Once the $\boldsymbol{\alpha}$ vector is known, computing α_0 is equivalent to finding the median of $\boldsymbol{\alpha} \cdot \boldsymbol{x}$, which can be done in linear time in the number of $\boldsymbol{x}$ values.

The complete algorithm is described recursively. For a set of points we describe how to build the root node of the subtree and how to split the points into two subsets.

(i) Compute the covariance matrix of all points.
(ii) Compute the eigenvector with the largest eigenvalue of the covariance matrix. (We call this vector $\boldsymbol{\alpha}$.) Normalize so that $|\boldsymbol{\alpha}| = 1$.

(iii) Compute the median of $\boldsymbol{\alpha}\boldsymbol{x}$ and call it α_0.

(iv) Split all points in two sets ($\boldsymbol{\alpha}\boldsymbol{x} \leq \alpha_0$ and $\boldsymbol{\alpha}\boldsymbol{x} > \alpha_0$).

(v) Build the two descendants recursively.

See the interactive exercise "NN-Tree."

The cost of building a NN tree

We will assume that the tree has m data points and that each point has n dimensions. The algorithm is recursive and has the following complexity for every root of a subtree of size m:

$$
\begin{aligned}
NN(m) = {}& m \cdot n^2 && \text{for building the covariance matrix} \\
& + n^3 && \text{for computing the eigenvalues and eigenvectors} \\
& + m \cdot n && \text{for finding the median of } \boldsymbol{\alpha}\boldsymbol{x}, \alpha_0 \\
& + 2NN(m/2) && \text{for building the left and right subtrees recursively} \\
= {}& \mathcal{O}\left(n^2 m \left(n + \log m\right)\right).
\end{aligned}
$$

5.4 Searching the NN tree

Given a point x by its n-dimensional coordinates, we normally want to find the k nearest neighbors (k is typically between 1 and 10). We first describe a function (written in pseudocode) which returns a set of neighbors which are not farther than ε away from the searched point x (Figure 5.9). We will use this function later to find the k nearest neighbors of x.

Algorithm

```
NNSearch( T:tree, x:point, epsilon:numeric )
r := {};
if type(T,Leaf) then
    # search sequentially in the leaf-bucket
    for w in bucket(T) do
        if |w-x| <= epsilon then
            r := r union w
        fi
    od;
else
    x0 := alpha(T)*x; # (each node has an alpha and an alpha0)
    if x0 - alpha0(T) <= epsilon then
```

```
                  # left tree has to be searched
                  r := NNSearch( left(T), x, epsilon )
              fi;
              if x0 - alpha0(T) >= -epsilon then
                  # right tree has to be searched
                  r := r union NNSearch( right(T), x, epsilon )
              fi;
          fi;
          return(r)
          end;
```

If $\boldsymbol{\alpha}(T) \cdot \boldsymbol{x} > \alpha_0(T) + \varepsilon$, we are at least a distance epsilon away from the left subtree. This allows us to discard the left subtree completely.

The mathematical deduction of this assertion poses no problem. For all $\boldsymbol{x}_i$ in the left subtree, i.e. $\boldsymbol{x}_i$ with $\boldsymbol{\alpha}(T) \cdot \boldsymbol{x}_i < \alpha_0$ follows

$$\boldsymbol{\alpha}(T) \cdot \boldsymbol{x} - \boldsymbol{\alpha}(T) \cdot \boldsymbol{x}_i > \varepsilon \text{ and hence } \boldsymbol{\alpha}(T) \cdot (\boldsymbol{x} - \boldsymbol{x}_i) > \varepsilon.$$

Since $\|\boldsymbol{\alpha}\| = 1$ this means $\|\boldsymbol{x} - \boldsymbol{x}_i\| > \varepsilon$ (using the Cauchy–Schwarz inequality $\|\boldsymbol{a} \cdot \boldsymbol{b}\| \leq \|\boldsymbol{a}\| \cdot \|\boldsymbol{b}\|$). Equality searching in the NN tree can be done by searching for NNSearch(T, x, 0).

Please see the interactive exercise "NNSearch."

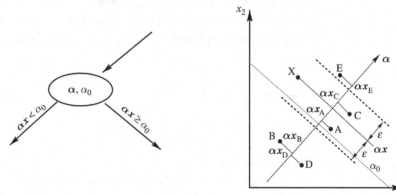

Figure 5.9   The function NNSearch.

5.4.1 Searching the k nearest neighbors

Searching the k nearest neighbors is done based on the ε-search described above. The algorithm works like this:

```
Search( T:tree, x:point, k:integer )
    eps := 1.e-5;
    # eps>0 should be set to a very small distance,
    # which will be normally satisfied by one data point.
    # Grow eps to include at least k points
    while true do
        t := NNSearch( T, x, eps );
        if size(t) < k then eps := 2*eps
        else break
        fi
    od;
    # Sort points in t according to distance to x
    t := sort( t, w -> |w-x| );
    return( t[1..k] )
end;
```

It is clear that the function NNSearch will work faster the smaller the ε is. If $\varepsilon = 0$ only one branch will be taken at each iteration and the amount of work is proportional to the height of the tree or $\mathcal{O}(\log m)$. For $\varepsilon > 0$ some nodes will require that we explore both descendants. The larger we choose ε, the more nodes there will be where we have to explore both descendants. In the limit, $\varepsilon \to \infty$, we will explore the entire tree ($\mathcal{O}(m)$ work).

So when we search for the k nearest neighbors we want to err on the smaller side for ε, not on the larger. That is why the Search algorithm is coded this way.

5.4.2 Concluding a value from its neighbors

This is usually called "the interpolation phase." We have several choices based on k, the number of closest points taken into account, and the type of problem.

(i) Search for $k = 1$ point (the closest point) and return its value.

(ii) Majority (in decision problems). Search for $k > 1$ next neighbors and make a majority decision in the following way:

$0 \ldots \frac{k}{3}$ values are "yes" $\Rightarrow$ return "no"

$\frac{k}{3} \ldots \frac{2}{3}k$ values are "yes" $\Rightarrow$ return "do not know"

$\frac{2}{3}k \ldots k$ values are "yes" $\Rightarrow$ return "yes."

(iii) Interpolation (in numerical problems). Search for k neighbors and return the average of the k neighboring values:

```
s := Search( T, x, k );
r := sum( w[value], w=s );
return( r/k )
```

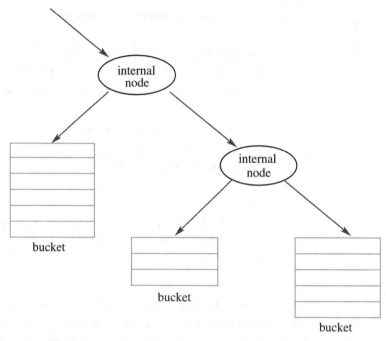

bucket

bucket

bucket

A subtree with buckets. Bucket size $b = 7$; for $b > 7$ the two buckets on the right would be merged, for $b < 7$ the one on the left would be split.

Alternatively we can compute a weighted average of the points. Let `half` be a distance derived from some other criteria for our set, such that points `half` away should be weighted $\frac{1}{2}$ relative to those at distance 0. An exponential weighted average is then computed with:

```
S := Search( T, x, k );
    wei := 0;
    sumv := 0;
    for w in S do
        dist := |w-x|;
        t := 2^(-dist/half);
        wei := wei + t;
        sumv := sumv + w[value] * t
    od;
return( sumv/wei );
```

A normally weighted average changes the assignment for t to

$$t := \exp\left(-\ln(2) * \left(\frac{\text{dist}}{\text{half}}\right)^2\right).$$

For numerical problems when the number of training data points is small, the above method is very convenient, as it has one parameter (the `half` distance) and is simple to code and understand. It does not require any neighbor computations, all nodes are used with their proper weight.

5.5 Practical considerations

5.5.1 Bucketing

This technique can be used to complement all tree methods. Bucketing is particularly useful, when the work (or space) needed for each internal node is relatively large. In this case, every time that a subtree is smaller than a certain threshold b, the subtree is not built. Instead all the values are stored in a bucket (Figure 5.10) which is just an array of leaf nodes. Any search in a bucket has to be done by searching all the records in the bucket, i.e. never more than b.

For the nearest neighbor trees, we usually choose $b \geq n$, so that the covariance matrices are not singular. (A covariance matrix of dimension $n \times n$ built from fewer than n data points will be singular.)

5.5.2 Normalizing dimensions

When we have dimensions which represent different magnitudes (or have different scales), the distances have to be weighted to make sense and be consistent. It is easy to see why this is needed. Just think of distances, some measured in kilometers others in millimeters; it will not make sense to add them as orthogonal measures without normalization. Instead of weighting the distances we can normalize each dimension so that the individual variables have average zero and variance one. This is sometimes called Z-transform or Z-score and is done by using

$$x' = \left(\frac{x_1 - \bar{x}_1}{\sigma_{x_1}}, \ldots, \frac{x_n - \bar{x}_n}{\sigma_{x_n}} \right),$$

where $\bar{x}_i$ is the average of dimension i and $\sigma_{x_i}^2$ is its variance.

5.5.3 Using random directions

An alternative to computing the covariance matrix and the eigenvalue/eigenvector decomposition is to try several random vectors α and choose the one for which

αx has the largest variance. This is only an approximation, but it is much easier to code:

```
# Using random directions to estimate alpha
# for a set of points of size m
var := -1;
to 20 do
    alpha := Random_Direction(n);
    x1 := x2 := 0;
    for x in set_of_points do
        t := alpha * x;
        x1 := x1 + t;
        x2 := x2 + t*t;
    od;
    t := x2*m - x1*x1;
    if var < t then var := t; best_alpha := alpha fi
od;
```

If we use k trials, constructing the NN tree will only cost $\mathcal{O}(kmn \log m)$ as opposed to $\mathcal{O}(n^2 m (n + \log m))$.

5.6 NN trees used for clustering

The NN tree is an ideal structure for clustering and cluster analysis. Clustering is understood to be the process by which we form disjoint subsets of an original set in such a way that the distances between elements of the same subset are shorter than the distances between elements of different subsets. A tree provides a way of defining clusters hierarchically, which is normally very appealing. By definition, any subtree can be a subset/cluster. The deeper we go in the tree, the more refined are the clusters, the higher we are in the tree, the more general are the clusters.

When NN trees are used for clustering, the following should be considered.

(i) Only meaningful dimensions should be kept. Useless dimensions (dimensions appearing to be random) will scatter the clusters and make their recognition more difficult. It is difficult to give criteria to select automatically good or bad dimensions for clustering. If a histogram of all points per dimension is computed, a good dimension for clustering will show as in Figure 5.11. A bad dimension for clustering will show as in Figure 5.12.

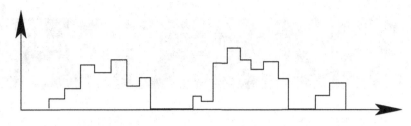

Figure 5.11   A good dimension for clustering.

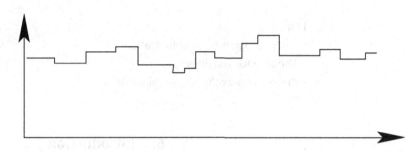

Figure 5.12   A bad dimension for clustering.

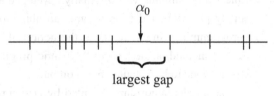

Figure 5.13   A good split at the largest gap.

(ii) The split (selection of α_0) should be done at a large gap, not necessarily at the median (see Figure 5.13). This could unbalance the trees, but it is the right criterion. If the balancing of the tree is more important than the quality of the clustering, one may select the largest gap such that the ratio of nodes on each side of the tree is not more than $k : 1$ with e.g. $k = 5$.

The clusters will be given by entire subtrees. Additional criteria are necessary to select these clusters (e.g. maximum distance between two points in the subtrees, maximum size, etc.).

Secondary structure prediction with linear programming (LP)

Topics
- Linear programming optimization
- Use of slack variables
- Convex hypervolumes, simplexes

6.1 Introduction

Linear programming was originally developed after the Second World War to solve supply problems and soon became an ubiquitous tool for optimization. The term "programming" in LP refers to the notion of optimization problems, solved by a tabular method, analogous to "dynamic programming" and not to the activity of writing code to be executed by a computer.

The simplex algorithm, invented by Dantzig in 1947 was the first very efficient tool for solving LP problems. Later, in 1972 Klee and Minty showed that the simplex algorithm would require exponential time for some pathological problems. It may also suffer from numerical instabilities for large problems. Nevertheless it is still an often used tool. During the 1980s, substantial contributions were made to LP, improving this situation. For example, Karmarkar's algorithm, introduced in 1984, is guaranteed always to finish in polynomial time and is numerically stable.

BASIC

Linear programming optimizes a linear function of the unknowns subject to linear constraints. Let $\mathbf{A}$ be an $m \times n$ matrix, $\boldsymbol{b}$ an m-dimensional vector and $\boldsymbol{x}$ our n-dimensional vector of unknowns. Normally $m > n$. The linear constraints in matrix notation are

$$\mathbf{A}\boldsymbol{x} \leq \boldsymbol{b}. \tag{6.1}$$

Let $\boldsymbol{c}$ be an n-dimensional vector, then the function to be optimized (maximized) is

$$f(\boldsymbol{x}) = \boldsymbol{c}\boldsymbol{x} = \sum_{i=1}^{n} c_i x_i.$$

This function f is usually called the *objective function*.

If we think in terms of an n-dimensional space, the solutions of a linear programming problem, if any exist, are points in the n-dimensional space. Each inequality of $Ax \leq b$, i.e. $A_i x \leq b_i$ is a *constraint* on the solution space and defines a hyperplane which divides the n-dimensional space in two parts: *feasible* and infeasible points. The intersection of all these half-spaces gives a convex subspace called a *simplex*.[1] If the simplex, i.e. the space defined by $Ax \leq b$, is empty, then the problem has no possible solution and is called *infeasible*. The determination of whether the simplex is empty or not is called the *feasibility problem* and is basically as difficult as LP itself.

6.1.1 An introductory example

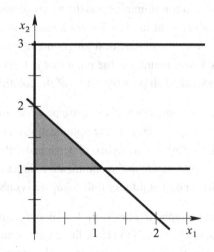

Figure 6.1   A simplex in two dimensions.

BASIC   Figure 6.1 shows a simplex in two dimensions (x_1, x_2) with $m = 4$, $n = 2$, given by the four conditions

$$x_1 \geq 0$$
$$x_2 \geq 1$$
$$x_1 + x_2 \leq 2$$
$$x_2 \leq 3.$$

---

[1] Note that *simplex* has two meanings: the algorithm invented by Dantzig and the subspace of feasible points.

In matrix notation this becomes:

$$\mathbf{A}x = \begin{pmatrix} -1 & 0 \\ 0 & -1 \\ 1 & 1 \\ 0 & 1 \end{pmatrix} \begin{pmatrix} x_1 \\ x_2 \end{pmatrix} \leq \begin{pmatrix} 0 \\ -1 \\ 2 \\ 3 \end{pmatrix} = \mathbf{b}.$$

In this case the simplex exists and is finite. If we remove the third constraint, $x_1 + x_2 \leq 2$, then the simplex also exists but is not finite, since it will extend to the right indefinitely. It is also easy to see that if we invert the fourth constraint, i.e. $x_2 > 3$, then the simplex does not exist and the problem, regardless of the objective function, is infeasible.

When the simplex is not empty, the optimization problem has a solution. The solution may not be unique and may not be finite. In the simplest case, if the simplex is of finite volume, then the solution is finite. In the above example, if $f(x) = x_2$ then the solution is unique, and the maximal point is $(x_1, x_2) = (0, 2)$.

It is useful to view the optimization as a direction c in the n-dimensional space. The solution of the linear programming problem is the point of the simplex for which cx is maximum, i.e. the point that is farthest in the direction of c. It is now easy to understand all possible forms of the solution:

(i)   a single point if the extreme edges of the simplex (in the c-direction) are not perpendicular to the optimization direction,
(ii) a line (or a higher dimensional subspace) if the extreme edges (faces) are orthogonal to the optimization direction,
(iii) a point at infinity if the simplex is unbounded in the direction of c.

For example, if $f(x) = x_1 + x_2$ in our example, then the solution is the line segment going from $(0, 2)$ to $(1, 1)$, finite but not unique, i.e. case (ii).

Finally if we remove the third constraint as we discussed before, the simplex becomes unbounded, it extends to the right infinitely. $f(x) = x_1 + x_2$ will have an infinite solution ($x_1 = \infty$, $x_2 \in [1, 3]$), i.e. case (iii).

It is also clear that if the simplex is not empty, is finite and $c \neq 0$, then the optimum will be at the boundaries of the simplex. It can never be in the interior as from any interior point we can move in the direction of c (improving f) until we hit the boundary. Having the optimal value at a boundary of the simplex is a consequence of $f(x)$ being linear in x and is a crucial property used by the simplex algorithm.

The constraints defining hyperplanes which are part of the simplex (i.e. are actually a face of the simplex) are called *active constraints*. In our example the first three constraints are active, the fourth, $x_2 \leq 3$, is inactive, it does not delimit the simplex. Consequently the optimum happens on some hyperplanes which are active constraints.

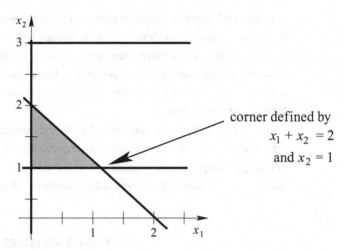

corner defined by
$x_1 + x_2 = 2$
and $x_2 = 1$

Figure 6.2   A corner in a simplex.

So, excluding case (iii) where the solution is unbounded, the optimal point lies on a boundary. And this means that we find the optimum at one or more corners – simply by inclusion, as in case (ii) the line or subspace with the extreme value also includes corners.

6.2 The simplex algorithm

BASIC

The simplex algorithm works by exploring the corners of the feasible region (Figure 6.2). It walks along the edges of the simplex from corner to corner, trying to improve the functional at every move. It is important to notice that a high-dimensional space may have an exponential number of corners, and any method based on the analysis of all corners is doomed to fail. This is not easy to understand by visualization as two-dimensional and three-dimensional spaces are simple. To understand this think of the n-dimensional hypercube which is defined by the $2n$ constraints, $x_i \leq 1$ and $x_i \geq 0$ for $i = 1, \ldots, n$. This hypercube has 2^n corners. So enumerating all the corners for $n > 40$ is out of the question even for the fastest current computers. The same applies to edges and faces in a high-dimensional simplex. This explosion of the number of corners is the main reason for the difficulty of LP. It is also the reason why the simplex algorithm moves from corner to corner rather than a more direct approach. A corner of the simplex is defined by n active constraints transformed into equalities, for example $A_2 x = b_2, A_5 x = b_5, \ldots$ (n equations). An edge is defined by $n - 1$ active constraints transformed into equalities. It follows quite easily that normally each corner is an end to n edges. (A corner is defined by n active constraints, by removing any one of them we obtain an edge which reaches the corner.)

So the simplex algorithm moves from a corner (n active constraints made equalities) through an edge (one of the n possible subsets containing $n - 1$ equalities) to a neighboring corner (n active constraints with only one different constraint/equation from the previous corner).

It is a very common assumption that all the variables are non-negative, as they reflect resources. This is not a significant restriction but should be remembered when defining the hyperplanes, number of inequalities and m. We see that if $m < n$, the simplex cannot have any corners and the solutions, if any, are infinite.

6.2.1 Slack variables

Finding any corner is a difficult problem, as difficult as LP itself. This is equivalent to finding a set of n inequalities out of $\binom{m}{n}$ which when turned into equalities define a point which satisfies all other $m - n$ inequalities. A general technique, introducing *slack variables*, is used for solving this and other similar problems. A slack variable is a new, artificial variable which should not be part of the final solution. For each unsatisfied inequality one slack variable is introduced to satisfy it.

To find a corner (and be able to use the simplex algorithm) we start at an arbitrary point (e.g. $x = 0$ for the cases when all $x_i \geq 0$) and add up to m slack variables – as many as necessary to transform all the unsatisfied inequalities into equalities, i.e. one slack variable for each $b_i < 0$.

In our example we need one slack variable $s = (s_1)$. In matrix notation this becomes

$$\mathbf{A} \cdot \begin{pmatrix} x \\ s \end{pmatrix} = \begin{pmatrix} -1 & 0 & 0 \\ 0 & -1 & -1 \\ 1 & 1 & 0 \\ 0 & 1 & 0 \end{pmatrix} \begin{pmatrix} x_1 \\ x_2 \\ s_1 \end{pmatrix} \leq \begin{pmatrix} 0 \\ -1 \\ 2 \\ 3 \end{pmatrix} = b.$$

The problem including the slack variables is called an augmented problem. What is useful about this augmented problem is that it defines a corner. This is true for the point $x = 0$, $s = 1$ (using all $x_i \geq 0$ and the unsatisfied inequalities turned to equations with the slack variables).

Normally, the simplex algorithm will first get rid of all the slack variables (and moves to a corner of the original simplex) and then does the real optimization. If we cannot optimize $f(x)$ any longer and have failed to remove all the slack variables (i.e. to make them zero), then the problem is not feasible.

We also have to remember that the slack variables are non-negative ($s \geq 0$) and that they are unwanted in the final solution, hence the functional to optimize has a very large negative cost associated with each slack variable, for example

$$f(x, s) = \sum c_i x_i - 10^{10} \sum s_i.$$

(We would like to use $-\infty$ instead of -10^{10}, but our computations should succeed in a real floating point system and hence it is much better to use a large constant.)

In our example the first and second row of A are the equalities

$$-x_1 = 0$$
$$-x_2 - s_1 = -1$$

which give us our starting corner $(x, s) = (0, 0, 1)$. We now proceed from corner to corner with this augmented problem, improving the functional at each step. If at any move one of the slack variables is made zero, then we immediately remove it from the solution, it does not need to be considered further.

When positioned in one corner, we can compute the direction of each edge. Each edge from a corner is defined by one missing equation of the corner. In our example one edge is given by removing the first equation, i.e. by

$$-x_2 - s_1 = -1 \quad \text{or} \quad x_2 + s_1 = 1$$

and the other edge by removing the second equation, leaving us

$$-x_1 = 0.$$

Let us use the first choice which is also simple to understand. We can move on a line which keeps $x_2 + s_1 = 1$ as long as all other inequalities are satisfied, i.e. $x_1 + x_2 \leq 2$, $x_2 \leq 3$, $s_1 \geq 0$. s_1 is at 1 and has a very large negative cost, because it is a slack variable, so we reduce it as much as possible to $s_1 = 0$ and so $x_2 = 1$.

At this point we can remove s_1 entirely from our problem, it has served its purpose; it took us from a corner in the augmented problem to a corner in our original problem. Our problem stands now as

$$\left. \begin{array}{r} -x_1 = 0 \\ -x_2 = -1 \end{array} \right\} \quad \text{defining a corner}$$

$$\left. \begin{array}{r} x_1 + x_2 \leq 2 \\ x_2 \leq 3 \end{array} \right\} \quad \text{other constraints}$$

$$\text{current corner} \quad (x_1, x_2) = (0, 1).$$

Let $f(x) = x_1 + 2x_2$ be our functional. The two edges coming out of our corner $(x_1, x_2) = (0, 1)$ are defined by

$$-x_2 = -1, \quad -x_1 \leq 0, \quad f(x) = x_1 + 2$$

and

$$-x_1 = 0, \quad -x_2 \leq -1, \quad f(x) = 2x_2.$$

For each edge we can compute how much the functional will increase, and we see that along both edges the functional increases.

Now we have to compute how far we can travel along the edge. This will be up to the point that the first inequality (outside of the active set) is violated. At this point we have reached a new (better) corner.

Suppose we take the first edge

$$x_2 = 1, \quad x_1 \geq 0,$$

then the other constraints become

$$x_1 + 1 \leq 2 \quad \text{and} \quad 1 \leq 3$$

which means that we can increase x_1 from zero to one, i.e. we move from one corner to the next one. The new corner is $(x_1, x_2) = (1, 1)$ and the problem is now defined by

$$\left. \begin{array}{c} -x_2 = -1 \\ x_1 + x_2 = 2 \end{array} \right\} \quad \text{define corner}$$

$$\left. \begin{array}{c} -x_1 \leq 0 \\ x_2 \leq 3 \end{array} \right\} \quad \text{other constraints}$$

$$\text{current corner} \quad (x_1, x_2) = (1, 1).$$

The next move will take us to the point $(0, 2)$ which is the solution of the problem.

When we are in a corner where no edge leads to an improvement, we have reached the optimum. This is the case with our example at $(0, 2)$, neither edge will improve the functional.

PRACTICAL NOTE Some algorithms define all unknowns to be ≥ 0. Others allow positive or negative values. These two modes are equivalent as we can transform each mode into the other.

- To solve a problem requiring non-negative variables with a solver that allows positive and negative variables, add $x_i \geq 0$ constraints to convert the second mode into the first one.
- To solve a problem using positive and negative variables with a solver that allows only non-negative variables, we can double the number of the unknowns and transform every x into $x_p - x_n$, where x_p is the positive value and x_n is the negative one, and hence convert the first mode into the second.

There are many details in the implementation and the handling of the simplex algorithm which are not covered here, we just want to give the main idea of the corner-walk.

6.3 SSP formulated as linear programming

The helix problem in secondary structure prediction attempted to find values for the variables (amino acid names) so that it would predict correctly where there are helices and where not. For example find the coefficients $A, C, D \ldots$ of the 20 amino acids plus one constant so that for all sliding windows around helices like in (3.2) we have

$$2A + D + T + G + c_0 \geq 1$$
$$\vdots$$

and for all windows around non-helices we have

$$R + 2A + D + T + c_0 \leq 0$$
$$\vdots$$

where the signal for a helix will be the linear form being greater than one and the signal for no helix is it being less than or equal to zero.

LP can be used in this problem for two goals:

- to determine feasibility,
- to optimize the linear coefficients for maximum reliability.

In our application we are mostly interested in determining feasibility. If we have inconsistent data, there is nothing to optimize. This case is discussed in Section 6.4.1.

If the problem is feasible we want the solution to be at the center of the simplex rather than at one corner, which is part of the boundary of the feasible region. Secondary structure prediction is a decision problem, and in decision problems the decisions happen at boundary points. So to have confidence in our solution, we should look for the feasible points which are as far away from the boundary as possible. It is easy to extend LP to find the center of a feasibility region, see Figure 6.3, next page.

First we should notice that given our problem, if we find a solution, multiplying the solution by an arbitrary factor larger than one will also give a solution, since all the equations will be multiplied by this factor and will still be true. Hence we have to place a restriction so that the simplex is finite. This is usually done by bounding each variable to reasonable values, for example

$$-1 \leq A \leq 1, \quad -1 \leq C \leq 1, \ldots$$

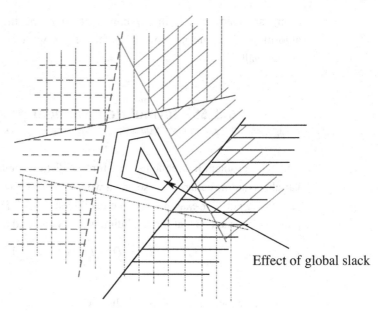

Effect of global slack

Figure 6.3   The effect of global slack.

on the basis that the numerical difference between the decision for helix or non-helix is one and a simple amino acid change alone could account for this difference.

In each inequality we add a new variable $Sl \geq 0$, called the global slack of the inequality:

$$2A + D + T + G + c_0 \geq 1 + Sl \qquad \text{(helix)}$$

$$\vdots$$

$$R + 2A + D + T + c_0 \leq 0 - Sl \qquad \text{(no helix)}.$$

The slack variable must be positive, so we must add an additional inequality:

$$Sl \geq 0.$$

Our functional is well defined now. We find the center of the feasible region by maximizing the global slack:

$$c \cdot x = Sl.$$

6.4 Inconsistent data

What if there is no feasible solution? This may happen because of

- inconsistent data or
- too many constraints, i.e. $n \ll m$.

There is one way to solve this problem: we have to sacrifice some inequalities. Sacrificing a minimum number of inequalities is a very difficult problem. So we will take simpler approaches.

6.4.1 Using slack variables for each inconsistency

Removing constraints to make the system feasible can also be done with slack variables. Let us call them S_i. We will add (or subtract) S_i to the inequalities which are inconsistent. (When a problem is inconsistent, most packages will return the inequalities which are being used, and hence contain the inconsistency.)

To make sure that we use the smallest number of slack variables possible we associate them with a large cost in the functional, say -10^6, or by a functional with only $-S_i$. For example, suppose that inequalities 1 and 2 belong to an infeasible set. Then we write

$$2A + D + T + G + S_1 \geq 1$$
$$S_1 \geq 0$$
$$R + 2A + D + T - S_2 \leq 0$$
$$S_2 \geq 0$$
$$c \cdot x = -S_1 - S_2.$$

The penalties for the S_i in the functional can also be used to reflect the confidence that we have in the training data. Suppose that we have two classes of data:

$$\text{class 1} \quad 2A + D + T + G + S_1 \geq 1 \quad \text{very good data}$$
$$\cdots$$
$$\text{class 2} \quad R + 2A + D + T - S_2 \leq 0 \quad \text{more doubtful data.}$$
$$\cdots$$

If we can quantify a relation between the errors of the classes, for example an error in class 1 is equivalent to twice the error in class 2, then we can implement this notion by setting in the c-vector $c_{S_i} = -1$ for the slack variables in class 1 and $c_{S_i} = -\frac{1}{2}$ for the slack variables in class 2.

6.4.2 Alternative method – removing inconsistencies

Another alternative is to assume that the data contained some errors, and we want to eliminate these errors. The problem of finding the minimum number of constraints to remove in order to find a solution is again very difficult (NP-complete). But the following greedy heuristic does quite well in general, although it is not guaranteed to find the best solution:

(i) assign all $c_{S_i} = -1$,
(ii) compute a solution,
(iii) eliminate the constraint which has the largest value of S_i,
(iv) repeat steps (i) to (iii) until a feasible solution (with all $S_i = 0$) is found.

6.5 Prediction

Once we have found a solution to the LP problem, call it α, the decision function used for prediction is:

$$\alpha_1 x_1 + \alpha_2 x_2 + \cdots \geq 1 \quad \text{(helix)}$$
$$\alpha_1 x_1 + \alpha_2 x_2 + \cdots \leq 0 \quad \text{(non-helix)}.$$

As in Section 4.7.1, we could ask the question whether an intermediate value could be better, for example using instead an arbitrary α_0 with

$$\alpha_1 x_1 + \alpha_2 x_2 + \cdots \geq \alpha_0 \quad \text{(helix)}$$
$$\alpha_1 x_1 + \alpha_2 x_2 + \cdots < \alpha_0 \quad \text{(non-helix)}$$
$$\text{and} \quad 0 < \alpha_0 < 1.$$

In the case of LP, an arbitrary α_0 does not help. If it would help, then there would be a better linear formula and this would be found by LP. Any constant between zero and one is equally good (based on the training data) to make the decision. We can settle conveniently for $\frac{1}{2}$, i.e.

$$\alpha_1 x_1 + \alpha_2 x_2 + \cdots \geq \frac{1}{2} \quad \text{(helix)}$$
$$\alpha_1 x_1 + \alpha_2 x_2 + \cdots < \frac{1}{2} \quad \text{(non-helix)}.$$

6.6 Comparison of linear programming with least squares

In LS   all the information is used: every data point "pushes" the solution in one direction.

In LP   only a subset of all data points are used.

- Some data points are just ignored because they already satisfy the constraints.
- Some are incompatible and discarded (no feasible solution).
- The unused ones (both of the above) play no role in the answer.

There is a good analogy between (LS, LP) and (mean, median). Given a set of values, the mean is the value which satisfies the condition of minimizing the sum of squares of its distance to each point in the set. The median is a value for which half of the values are less and half of the values are greater.

6.7 Example

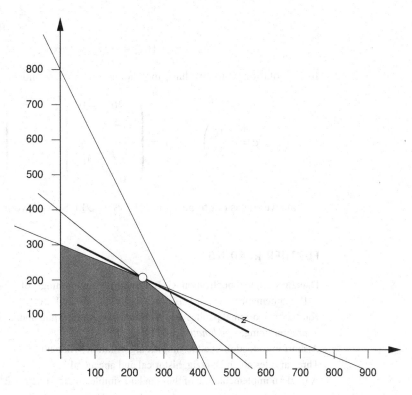

Figure 6.4   Graph of the linear programming example.

Here is a specific example of a problem in linear programming (see Figure 6.4).

	Women's shoes	Men's shoes	Available
production time [h]	20	10	8000
machine time [h]	4	5	2000
leather usage [dm^2]	6	15	4500
net income [$]	16	32	—

We choose the following unknowns: x_1 is the quantity of women's shoes produced; x_2 is the quantity of men's shoes produced.

The mathematical problem looks like this:

$$
\begin{aligned}
y_1 &= -20x_1 - 10x_2 + 8000 &\geq 0 \\
y_2 &= -4x_1 - 5x_2 + 2000 &\geq 0 \\
y_3 &= -6x_1 - 15x_2 + 4500 &\geq 0 \\
x_1 & &\geq 0 \\
x_2 & &\geq 0 \\
z &= 16x_1 + 32x_2 \overset{!}{=} \max.
\end{aligned}
$$

In the notation which we have introduced this corresponds to:

$$
c = \begin{pmatrix} 16 \\ 32 \end{pmatrix} \qquad
A = \begin{pmatrix} -20 & -10 \\ -4 & -5 \\ -6 & -15 \\ 1 & 0 \\ 0 & 1 \end{pmatrix} \qquad
b = \begin{pmatrix} -8000 \\ -2000 \\ -4500 \\ 0 \\ 0 \end{pmatrix}.
$$

The answer turns out to be $x_1 = 250$, $x_2 = 200$. Solve it yourself!

FURTHER READING

Dantzig's original publication: G. B. Dantzig, Programming of interdependent activities. II. mathematical model, *Econometrica*, **17** (1949) 200–211.

Karmarkar's original publication: N. Karmarker, A new polynomial-time algorithm for linear programming, *Combinatorica*, **4** (1984) 373–395.

A short introduction into the simplex algorithm in Wikipedia.[2]

Three articles about Dantzig, historical and anecdotal.[3]

A modern implementation of the standard simplex method: retroLP.[4]

---

[2] Wikipedia: http://en.wikipedia.org/wiki/Simplex_algorithm
[3] www.stanford.edu/dept/news/pr_stage/2005/pr-dantzigobit-052505.html
www.stanford.edu/group/SOL/dantzig.html
www.thehindubusinessline.com/bline/2005/06/24/stories/
2005062400630900.htm
[4] retroLP: http://cis.poly.edu/tr/tr-cis-2001-05.htm

A modern implementation of Karmarkar's algorithm.[5]

The simplex algorithm: W. H. Press, B. P. Flannery, S. A. Teukolsky and W. T. Vetterling, *Numerical Recipes in C: the Art of Scientific Computing*, Cambridge University Press, 1992, Chapter 10.8.

V. Klee and G. J. Minty. How good is the simplex algorithm? in O. Shisha, ed. *Inequalities III*, New York, Academic Press, 1972, pp. 159–175.

---

[5] Karmarkar's algorithm: `http://citeseer.ist.psu.edu/gay87pictures.html`

Stock market prediction

Topics
- Dynamic programming
- Optimization of decision-based functionals
- Simulation: modelling, validation, prediction

7.1 Introduction

BASIC

Stock market prediction is a very interesting topic and could also be very profitable if done successfully. It falls well within our definition of modelling/prediction. We define "stock market prediction" as the techniques which extract information from past behavior to decide when to buy or sell which stock and at what price. This is called quantitative analysis or technical analysis, i.e. the use of mathematics to predict future behavior based on historical data. It is normally contrasted with fundamental analysis which analyses the health of the business in a very broad sense to predict the future behavior of stock prices.

There are several reasons which make technical analysis a very good topic for the study of methods in modelling and prediction.

(i) The problem is simple and easy to define precisely.
(ii) All the main concepts of modelling/prediction appear in this problem.
(iii) There are abundant data – most publicly available through the World Wide Web.
(iv) New data are generated every day, which allows extensive and realistic validation.

Please note that the goal of this chapter is to illustrate methods in modelling and simulation. It is not a complete or up-to-date treatise on stock market prediction.

7.1.1 Definitions

We are going to use only buying/selling of regular stock and avoid more sophisticated tools such as options, futures etc. A few definitions are necessary to clarify the concepts.

Stock market An institution which trades stock (shares) for people who submit orders. For each company traded, a "book" is kept.

Bid   The price per share offered by someone trying to buy shares.

Bidsize   Number of shares, for which a buying offer is made at the bid price.

Ask   The price per share wanted by someone trying to sell shares.

Asksize   The number of shares, for which a selling offer is made at the ask price.

Spread   The difference between lowest ask and highest bid price (non-negative).

Book   A group of bids and asks and their associated sizes. A book is normally ordered by decreasing price for bids and increasing price for asks. The highest bid is usually called the current bid, the lowest ask is usually called the current ask. When the bid price is equal to or exceeds the ask price, i.e. $Spread \le 0$, transactions take place immediately. Here is an example of a book for some stock:

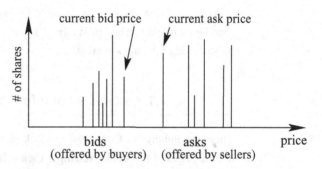

Open   The price paid at the first transaction of the day.

Close   The price paid at the last transaction of the day.

High   The maximal price paid during one day.

Low   The minimal price paid during one day.

Volume   The number of shares transacted during one day.

Transactions Every time a market order comes (and the book is not empty) or the bid price matches or exceeds the ask price, a (buy–sale) transaction will happen.

Order   The instruction to buy or sell a given number of shares.

Market order   The instruction to buy or sell a given number of shares at the current price, i.e. buy/sell at current ask/bid. Note that when buying a large number of shares with a market order, the bid stack can be exhausted and the price can get very high. Likewise when selling a large number of shares, the ask stack can be emptied and the price can drop a lot. There is no price limit for a market order!

Limit order   The instruction to buy a given number of shares when the price becomes lower or equal to the given limit, or sell a given number of shares when the price becomes higher than or equal to the given limit. There is no guarantee that such an order will be executed, it will just be added to the corresponding stack.

Stop order   The instruction to buy or sell a given number of shares when the price of the stock reaches the stop value. Stop buy activates when the price is higher than or equal to the stop value, stop sell activates when the price is lower than or equal to the stop value. Stop orders behave in the opposite way to limit orders.[1]

Ticker symbol   An abbreviation, consisting of a few letters usually in capitals, which identifies a particular company or stock.

There are many other terms and concepts that we will not use in our examples. We will restrict ourselves to a very simple model of buying and selling stock.

At www.investopedia.com/dictionary/ you can look up some thousand stock market terms. In Wikipedia you can find a thorough general introduction by searching for "stock market."

7.1.2 Examples of information available online

There are hundreds of sites giving stock market information. Some essential pages are the pages of the stock exchanges themselves:

New York Stock Exchange	www.nyse.com/
NASDAQ	www.nasdaq.com/
London Stock Exchange	www.londonstockexchange.com/en-gb/
Euronext	www.euronext.com/
Frankfurter Börse	deutsche-boerse.com/

---

[1] This appears to be against logic (sell when the price hits a low barrier) so it requires an explanation. Suppose your stock has reached a price where you can sell it at a profit, say at $10.00 per share. But it may continue to go up. Or go down and diminish your profit. So you place a stop order at $9.50 per share. If it starts going down, you want to sell without a loss. If it keeps going up, you may increase the stop price.

Then there are dozens of independent portal pages giving stock information, for example

Onvista   www.onvista.com/
CNN   money.cnn.com/
Yahoo   finance.yahoo.com/

or you can try the stock portal pages of nearly every banking house, especially all the online brokers. Just search for "online broker" to get a selection of links. Finally, there are streamers showing the actual transaction prices in real time, for example streamer.com/ or www.wallstreet-online.de.

7.1.3 Rationale for stock market prediction

There are several theories about how the stock market behaves which are too specific to be covered here. These theories range from the market behaving like a purely random walk (i.e. completely unpredictable) to "full knowledge" or "efficient market hypothesis" theories where the market knows all the information (although we may not know all of it) and acts accordingly. We will not discuss these theories, we will simply discard the purely random walk theory, since such a theory precludes any prediction.

To illustrate the prediction process we will use an analogy. The analogy is that of a biased die. Let us assume we have a die which is biased, that is, the outcomes of rolling it are not equiprobable. (It could be e.g. $\Pr(1) = \frac{1}{4}$, $\Pr(2) = \frac{1}{2}$, $\Pr(3) = \Pr(4) = \Pr(5) = \Pr(6) = \frac{1}{16}$.) The bias of the die is unknown to us, all that we can observe is the outcome of several rollings, for example ... 1, 5, 1, 2, 3, 6, 1, 5, ... , 1. At this point we are asked to predict the next outcome and beat the odds (i.e. predict with a probability of success better than $\frac{1}{6}$). For this we observe the sequence of past results and attempt to guess the probability distribution. With this approximation we choose the number which has the highest probability of outcome (1 for the example sequence).

The market is analogous to the die and all its information. We do not know this exact information, but we can observe the past behavior (in the market, the sequence of prices and volumes; with the die, the sequence of outcomes). From these observations we try to extract information that may reveal the biases (of the market or of the die). To "guess" the behaviors we propose a model. (In the case of the die we assume an individual probability for each outcome.) For the stock market more complicated models will be proposed. Next we use the past behavior to approximate the parameters of our model (e.g. compute the probability distribution of the die). Finally we use this information to predict the most likely output of the next rolling.

Notice that if the die is biased and if the model is correct, then if we have enough data, we will indeed approximate the real probabilities and predict more accurately the next outcome. Exactly the same should be true with stock market prediction.

7.1.4 Organization of the modelling

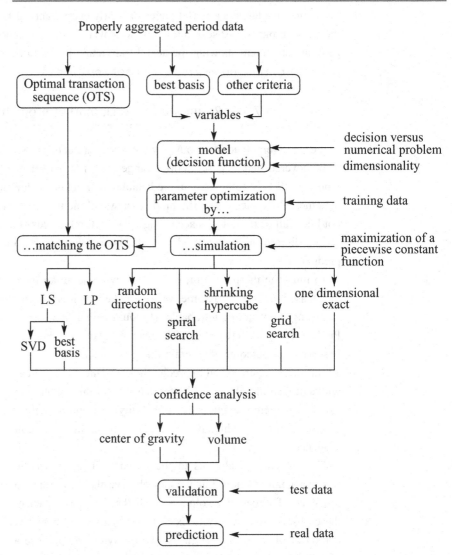

Figure 7.1 The general scheme of stock market prediction. The topics mentioned here will be developed in this chapter.

Figure 7.1 shows the various ways which we use to approach the stock market prediction problem.

The three main topics of this chapter are the computation of the optimal transaction sequence (OTS), the approximations to the OTS and simulation. The details of Figure 7.1 will become clear in the next sections.

7.2 The optimal transaction sequence (OTS)

7.2.1 Defining the problem

The first problem which we will solve is that of determining an optimal sequence of transactions between various stocks. This will be done knowing all the prices of the stocks, so it is not a prediction, it is just an a posteriori analysis of the best sequence of transactions that could have been made. We will simplify the problem as follows.

- There is only (at most) one transaction per day.
- We define a **fixed** spread of 0 between bid price and ask price. (There is always a spread > 0, i.e. bid price < ask price, otherwise a transaction happens, but we will ignore this detail.)
- Partial transactions do not give any benefit, so we do not consider them. (A partial transaction is one where we only use part of our capital to buy shares or we divide our investment between several stocks. Since we are interested in the *optimal* strategy, there is never any benefit in buying other stocks than the optimal.)
- No commissions or transaction costs will be charged.
- We use fractional numbers of shares (in reality we normally buy whole shares or even in some cases multiples of 100).

To solve this problem we use dynamic programming. For details see Section 7.6.

7.2.2 Calculating the OTS for the past

For the optimal transaction problem, every day will represent one period of time. Let k be the number of different stocks. The state for every day is one of $k + 1$ possibilities: k possibilities for having all the investment in stock number $1, \ldots, k$, the $(k + 1)$st possibility for having no stocks, only money. The possible transitions between one period and the next are always between the following states.

From state i to state i: this corresponds to doing nothing $(i = 1, \ldots, k + 1)$.
From state $k + 1$ to state i: buy stock of company i $(i = 1, \ldots, k)$ at the buying price for that day.

From state i to state $k + 1$: sell stock of company i ($i = 1, \ldots, k$) at the selling price
    for that day.

Each state will be associated with a value. For the states $1, \ldots, k$ this is the number
of shares, for state $k + 1$ this is the number of dollars. (See the following section for
an example.)

To select the price several choices are possible. We could take the price at a given
time in the day (e.g. *Open*) and buy at *Open + Gap* and sell at *Open − Gap* (*Gap*
being a small safety margin to insure that the transactions will happen). Alternatively,
we could take the optimistic (unrealistic) view that we could buy at *Low + Gap* and
sell at *High − Gap*. Or take a pessimistic view that we buy at *High − Gap* and sell at
Low + Gap (unrealistic but safest). Once this decision is taken, we will have a price
for each stock/transition and we can compute the volume after each transition. Of all
the possible transactions which lead to the table entry, we choose the one which gives
the highest value and record this value.

The optimal profit is given by state $k + 1$ of the last day.

The first row (see Figure 7.2) just states the initial condition, i.e. no stocks and $1000
in money. The following rows are computed from the previous rows. The optimal
gain is 154.32 (15.432%) in 8 trading days. The optimal sequence of transactions
is computed by backtracking from the bottom rightmost entry (the optimal path is
shown by the arrows, i.e. first day buy BHI, second day do nothing, third day sell
BHI, fourth day buy AVP etc.).

In the interactive exercise "Optimal transaction sequence" (see next section) you
can test this yourself. Try the shares of different companies.

7.2.3 Interactive exercise "Optimal transaction sequence"

Instructions for using the interactive exercise The interactive exercise shows the
prices of three stocks at eight consecutive periods of time.

- When you click the "Run algorithm" button, the maximal amounts of stock and
  money which can be reached at these periods will be calculated.
- When you press the "Show best transaction sequence" button, green arrows indicate
  by which operation these maximal values in the "Shares" and "Money" columns
  were reached in the forward part of the dynamics programming algorithm.
  A vertical arrow means: Nothing was done.
  An arrow from shares to cash means: Shares were sold.
  An arrow from cash to shares means: Shares were bought.
- When you press the "Show optimal path" button, red arrows and fields show the
  result of the backtracking part of the dynamic programming algorithm.

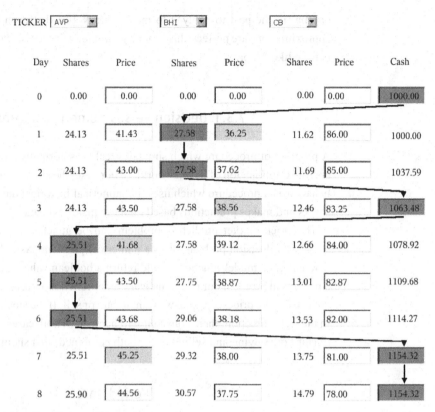

Figure 7.2   An optimal transaction sequence.

- When you click onto one of the arrows all the calculations made for choosing this transaction as the best possible are shown. (So you can compare the results of all possible transactions and see why the chosen one is optimal.)

Please use the online version of this interactive exercise.

For details on the algorithmic background of dynamic programming see Section 7.6.

7.3 Approximating the OTS

The OTS would have given us an optimal gain if we had used it with knowledge of the future. (It is very easy to become wealthy if you can predict the future accurately!) The main assumption regarding stock market prediction is that patterns of the past will be reproduced in the future, or in other words, that under similar conditions, prices will move in the same way. Hence we try to learn from what would have been optimal in the past to apply it to the future. Notice that even if our mathematical approximations are perfect, this is not a guarantee of success since the future may not behave like the past.

7.3.1 Decision versus numerical approach

A program or procedure which simulates reality is normally based on an abstraction and/or simplification of the real world. In our case, the simulation of the stock market behavior is a procedure which uses the numerical historical data of a stock or set of stocks and makes predictions based on these historical data.

The stock market prediction problem can be attacked in at least two different ways, as a decision problem or as a numerical problem. As a decision problem our mathematical model/function should return a boolean value (`true` or `false`). As a numerical problem our mathematical model/function returns a value. These values are usually the prices at which we want to buy or sell. If the stock reaches such a price in the right direction, then a buy/sell will happen. In both cases we will define values d_i (for both buying and selling or for both combined) that should be our goal.

Period data

A period i is a unit of time in which we observe variables of the stock and in which we will do transactions. A period normally has *Open*, *High*, *Low*, *Close* prices and a *Volume* of shares traded associated with it. A period could be for example a year, a day or an hour.

Period data are *any* data that are derived from the basic period information, i.e. *Open*, *High*, *Low*, *Close*, *Volume* etc. We will index these data with the period number i, for example $Open_i$ is the open price of period i, $Close_{i-2}$ is the closing price of period $i - 2$. For obvious reasons we are allowed to use $Open_i$ as it is available at the very beginning of period i, but neither $High_i$, Low_i, $Close_i$, $Volume_i$ for any decision made for period i. We can use the information of variable x_{i-k} for any variable x and any positive k: for example $Open_i$ the open price of period i, $Volume_{i-1}$ the volume in the period preceding i, but not $Close_i$ or $Volume_{i+1}$.

Fractal nature

Most stock data variables show fractal behavior. That is, if you plot them without labels, it is impossible to decide what is the time scale. This is an assumption, a quite realistic one, but one which cannot be proved. Compare the price graphs of the DAX

Figure 7.3   Development of the DAX over five days compared with the development over two years.

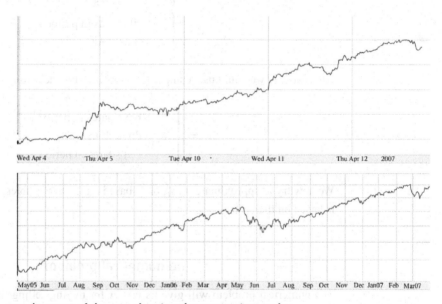

Wed Apr 4   Thu Apr 5   Tue Apr 10   •   Wed Apr 11   Thu Apr 12   2007

May05 Jun Jul Aug Sep Oct Nov Dec Jan06 Feb Mar Apr May Jun Jul Aug Sep Oct Nov Dec Jan07 Feb Mar07

Figure 7.4   Development of the DAX showing the appropriate scales.

index in Figure 7.3. One graph shows the development of the DAX over five days, the other the development over two years. Can you tell which is which?

Besides the actual slope, it is not possible to identify a pattern which tells whether they are at the same time scale or which one is larger. If the labels are not shown, it may be surprising to learn that the first graph is for five days, whereas the second one is for two years (Figure 7.4).[2]

---

[2] If the period is too short, e.g. one second, there may be so few transactions that the graph will look different. So we have to exclude very short periods for the fractal assumption.

We use this assumption of a fractal nature to our advantage. Our models do not have to be specialized for a particular period length, they can work with any time scale as long as we just use the variables *Open*, *High*, *Low*, *Close* and *Volume* per period.

The decision problem

The first approach to do our modelling is as a decision problem, i.e. we are going to construct two functions (one for buying and one for selling) returning the values b_i and s_i, which will determine *when* we will buy and sell:

$$b_i = \begin{cases} 1 & \text{buy shares in period } i \\ 0 & \text{do nothing} \end{cases}$$

$$s_i = \begin{cases} 1 & \text{sell shares in period } i \\ 0 & \text{do nothing.} \end{cases}$$

Alternatively we could use a single function with two decision points,

$$a_i = \begin{cases} 1 & \text{buy shares in period } i \\ 0 & \text{do nothing} \\ -1 & \text{sell shares in period } i. \end{cases}$$

We now try to approximate b_i, s_i, or a_i based on the x_i variables. The approximation is based on the OTS.

The numerical problem

The numerical problem will give us *values* for buying/selling the stock. In the numerical model we construct two functions giving values b_i and s_i for each period i. If the price falls below b_i we buy the stock in period i, and if the price climbs above s_i we sell the stock in period i.

The predictions will be executed perfectly if we manage to construct b_i and s_i to attain the following values:

$$b_i = \begin{cases} Low_i + Spread & \text{buy shares in period } i \\ \text{a value } < Low_i & \text{do not buy shares in period } i \end{cases}$$

$$s_i = \begin{cases} High_i - Spread & \text{sell shares in period } i \\ \text{a value } > High_i & \text{do not sell shares in period } i. \end{cases}$$

The information about buying or selling or doing nothing is determined by the optimal transaction sequence.[3]

So in both approaches we are left with the task of constructing mathematical functions which take period data as variables.

Constructing mathematical functions

The functions that will be used for making the decisions are mathematical functions which depend on the period data and on parameters. The period data are a subset of the data available at period i. The parameters (if any) are constants which are computed to optimize the results of the model based on the training data. An example of such a function would be

$$b_i = \frac{a_0 \cdot Open_i + a_1 \cdot Close_{i-1}}{High_{i-2} + a_2 \cdot Low_{i-3}}.$$

This function can be applied to any period (provided that we have data for at least three previous periods) and it has three parameters, a_0, a_1, and a_2. Computing the best/optimal values of the parameters is called training and will be discussed in the next section.

First let us discuss, which period data variables to use. Besides the already mentioned period data variables we can compute

- averages,
- max/min,
- variance and standard deviation,
- linear or quadratic interpolations.

Any of these can be computed over any of *Open/High/Low/Close/Volume* and over any past period, for example

$$\max_{k=0\ldots10}(Open_{i-k})$$

$$\text{Stddev}_{k=1\ldots100}(Volume_{i-k})$$

$$\text{Average}_{k=10\ldots20}(High_{i-k} - Low_{i-k}).$$

Another example would be to use the coefficient a_1 of a linear least squares approximation over some range, for example

$$a_0 + a_1 j \approx Open_{i-j} \quad \text{for} \quad j = 0\ldots10.$$

---

[3] Buying and selling are obviously similar but opposite actions. The extent of the similarity goes beyond first appearances, as we can sell stock that we do not necessarily have (short sell). In summary, if we have capital, regardless of whether we have the stock or not, we can buy or sell. (Selling short means borrowing a stock you do not have and selling it. It is up to the person to return the stock later.)

In this case we capture how the opening price increases or decreases as a function of time. Another example of a function would be the coefficient a_2 in the quadratic approximation of closing prices:

$$a_0 + a_1 j + a_2 j^2 \approx Close_{i-j} \quad \text{for} \quad j = 0 \ldots 20,$$

which will give the orientation of the parabola that fits the closing prices. If $a_2 > 0$ then the parabola is upwards and we should expect the prices to have bottomed and start climbing. If $a_2 < 0$ the prices have peaked and will start decreasing.

It is obvious that there is a tremendous amount of choice for the variables to be used. It is impossible to try any reasonable number of subsets. So we use statistical techniques (based on least squares) to select the variables which are best correlated. The technique, illustrated for the case of a decision problem, is the following.

- Use d_i, the results of the optimal transaction sequence.
- Use a linear (in the parameters) model l_i involving all variables (or as many as it seems reasonable or possible to use). Make it dimensionally consistent (see the following subsection).
- Use best basis to select a small subset $l_k^*(i)$ of variables which contribute the most to explain the decisions: $\sum \alpha_k l_k^*(i) \approx d_i$.
- Use the variables selected from best basis to write a function. For example, variables with positive coefficients could be in the numerator, variables with negative coefficients in the denominator,

$$l_i = a_1 Open_i + a_2 High_{i-2} + a_3 Low_{i-1} + a_4 Close_{i-1} + \frac{a_5}{Vol_{i-1}} + \cdots .$$

If best basis of size 4 returns

$$l_4^*(i) = 0.3 Open_i - 0.2 Close_{i-1} + 0.15 High_{i-2} - 0.2 Low_{i-1}$$

then a possible dimensionless formula would be

$$\frac{a_1 Open_i + a_2 High_{i-2}}{a_3 Low_{i-1} + a_4 Close_{i-1}}.$$

Dimensionality

Almost any formula is possible. There is, however, a dimensionality criterion to be respected. Stocks may split (e.g. one share is divided into two, the price of the shares halves and the number of shares doubles) depending on absolutely arbitrary reasons. (A company decides to split stocks based on the local perception of prices and possible market restrictions, for example North American markets like prices in the range $10–$100 per share, European markets like higher values per share.) A stock split is inconsequential for the shareholder's value. So a good formula should be consistent under a stock split. That is, the predictions before and after the split should be identical.

Under a stock split of factor f every price is divided by f, and every volume is multiplied by f. f is not necessarily an integer although it is usually a small fraction. If the formula is intended to predict a price it should have the dimensionality of price, if it is intended to be used in a decision it should be dimensionless, for example

$$\frac{a_0 Var_{k=1\ldots 10}(High_{i-k}) \cdot Vol_i + a_1 Open}{High_{i-1} + a_2 Low_{i-2}}.$$

This formula is dimensionless and hence suitable for a decision. Under a split of factor f all terms in the numerator and denominator will be multiplied by $1/f$, so the final value is the same.

The formula $Stddev_{k=1\ldots 5}(Close_{i-k}) + a_0 High_{i-1} + a_1$ is dimensionally not correct, since two terms are multiplied by $1/f$ and the last is not affected. To check the dimensional consistency we use the following simplification. We replace each value for the factor after the split. The products/quotients are simplified and every partial sum in the expression should have only terms with the same power of f. The first formula becomes

$$\frac{\frac{1}{f^2} \cdot f + \frac{1}{f}}{\frac{1}{f} + \frac{1}{f}} = 1.$$

Both sums, numerator and denominator, have consistent powers of f. Since this is a constant (independent of f), the formula is dimensionally consistent for a decision. If it were a constant times $1/f$ it would be suitable for a formula giving a price value. The second formula above would give $1/f + 1/f + 1$ which is neither a constant nor a factor of $1/f$, hence it is dimensionally not uniform.

7.3.2 Computing good (optimal) parameters

Once we have the model (i.e. our mathematical function and decision method) we can use the historical data as training data to select the best values for the parameters: we fit the model to the data. Recall the standard training/validation/prediction diagram in Figure 3.3. To give an overview again, Figure 7.5, next page shows the general scheme adapted and detailed for stock market prediction.

As we said before, there are two main approaches to solving this problem. The first is based on the optimal transaction sequence and the second is based on simulation. The problem of approximating the optimal transaction sequence becomes almost identical to the problems described in Chapters 3–6 for α-helix prediction. We can

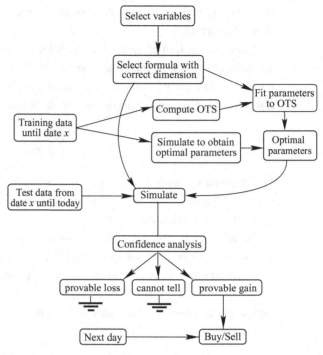

Figure 7.5   The general scheme of stock market prediction.

use least squares to approximate the decisions, either using SVD or best basis. We can also use linear programming to try to satisfy as many optimal decisions as possible.

For numerical problems it is also very similar. In particular, let $b(x_i)$ be a buying function and $s(x_i)$ a selling function. Let $d_i = 0$ (no action), $d_i = 1$ (buy) and $d_i = -1$ (sell) be the optimal decisions from the optimal transaction series.

The least squares equations are

if $d_i = 0$	$s(x_i) \approx High_i + g$	(never reached)
(do nothing)	$b(x_i) \approx Low_i - g$	(never reached)
if $d_i = 1$	$b(x_i) \approx Low_i + g$	(reached at a good buying price)
(buy)	$s(x_i) \approx High_i + g$	(never reached)
if $d_i = -1$	$b(x_i) \approx Low_i - g$	(never reached)
(sell)	$s(x_i) \approx High_i - g$	(reached at a good selling price)

where g is a gap value that will insure that transactions happen or do not happen with some safety margin.

If instead of a least squares approximation we use linear programming, the inequalities are quite similar:

if $d_i = 0$	$s(x_i) > High_i$	(never reached)
(do nothing)	$b(x_i) < Low_i$	(never reached)
if $d_i = 1$	$Low_i + g \geq b(x_i) \geq Low_i$	(reached at a good buying price)
(buy)	$s(x_i) > High_i$	(never reached)
if $d_i = -1$	$b(x_i) < Low_i$	(never reached)
(sell)	$High_i - g \leq s(x_i) \leq High_i$	(reached at a good selling price)

If we are solving this problem as a linear programming problem, a feasible solution means that we would follow the optimal transaction sequence. The variable g allows a safety margin for transactions to happen at a less than optimal price. Hence a big value for g will give poor results. In LP we can maximize the functional $c \cdot x = -g$ so that we minimize g and hence trade as close to the optimum as possible.

7.3.3 From the OTS to reality

So far we have optimized the approximations, i.e. we have found functions $b(\ldots)$, $s(\ldots)$ which are as near as possible to the optimal values, following the optimal transaction sequence. This approximation is not necessarily perfect, and even with the best approximation some decisions may be wrong. With the above approximations we have no control over whether an incorrect decision will have a minor or major impact on the final gain.

A second approach, the one developed for simulation in Section 7.4, considers the gain/loss of a given model and attempts to maximize this gain, as opposed to making each single decision optimal, i.e. we find the parameters for which the total gain is maximal.

The goal of optimizing towards the OTS is the best that we could hope for. But since this approximation is seldom perfect, large deviations may occur.

7.4 Simulation

Simulation is one of the most important applications of computer science. It allows particular environments or situations to be recreated and the result of applying policies or design criteria to be tested. As computers get faster and larger we are able to simulate reality more and more accurately. Simulation is used in a very wide range of applications, for example wing design in airplanes, national economic policies, weather prediction, evolution of diseases, warming of the atmosphere ... just to name a few.

In our example, the stock market, simulation is a very powerful tool and it is quite easy to use. It allows us to evaluate models (buy/sell strategies) using historical data.

7.4.1 Two simple models

As examples we look at two strategies (both are quite naive).

Strategy 1

- Buy with a limit order when the price falls below the opening price by some amount p.
- Sell with a limit order when the price goes above the opening price by the same amount p.

Using limit orders means that the buy/sell, if it happens, will happen exactly at that price. We consider this transaction done when the price goes slightly (i.e. by gap) beyond the limit price. We run the simulation with the prices for the stock of Yahoo, a company trading on the Nasdaq stock exchange. We use all the year 2005 for this run.

```
# Simulation (strategy 1)
S := 0:        # stands for the number of shares
M := 1000:     # stands for the capital (money)
p := 1:        # price increase/decrease to sell/buy
gap := 0.05:   # gap to guarantee the transaction happens

for i from 1 to length(Price) do
    if M>0 and Price[i,OPEN]-p >= Price[i,LOW]+gap then
            S := M / (Price[i,OPEN]-p);
            printf( 'on %d buy %.3f sh at %.2f \n',
                    Price[i,DATE],S, Price[i,OPEN]-p );
        M := 0
    elif S>0 and Price[i,OPEN]+p <= Price[i,HIGH]-gap then
            M := S * (Price[i,OPEN]+p);
            printf( 'on %d sell %.3f sh at %.2f, M=%.2f \n',
                    Price[i,DATE], S, Price[i,OPEN]+p, M );
            S := 0
    fi;
od;
# if there are any shares, value them at the close of the
# last day
```

```
M := M + S*Price[length(Price),CLOSE]:
printf( 'final capital: $ %.2f\n', M );
```

```
    on 20050104 buy 26.702 sh at 37.45
    on 20050211 sell 26.702 sh at 34.45, M=919.89
    on 20050223 buy 28.909 sh at 31.82
    on 20050224 sell 28.909 sh at 31.43, M=908.62
    on 20050314 buy 29.558 sh at 30.74
    on 20050330 sell 29.558 sh at 33.31, M=984.58
    on 20050607 buy 26.102 sh at 37.72
    on 20050929 sell 26.102 sh at 33.40, M=871.82
    on 20051129 buy 21.790 sh at 40.01
    on 20051130 sell 21.790 sh at 40.38, M=879.88
    on 20051219 buy 21.377 sh at 41.16
    final capital: $ 837.56
```

Since the initial capital was $1000 we would have lost about 16% of our money if we applied this strategy during the year 2005 to the stock of Yahoo. Not a very promising result.

Notice that the program follows exactly the policy stated above. This is the main characteristic of simulation, we recreate our actions over some data (in this case real historical data) with the purpose of evaluating the behavior of our model.

Strategy 2

- Sell when we have shares and the price starts to fall, that is when the opening price has fallen more than $p\%$ compared with the previous close.
- Buy when we have capital and the price starts to climb, that is when the opening price has risen more than $p\%$ compared with the previous close.

```
# Simulation (strategy 2)
S := 0:        # stands for the number of shares
M := 1000:     # stands for the capital (money)
p := 2:        # percentage fall/rise to sell/buy

for i from 2 to length(Price) do
    if M>0 and Price[i,OPEN] > (1+p/100)*Price[i-1,CLOSE] then
        S := M / Price[i,OPEN];
        printf( 'on %d buy %.3f at %.3f\n',
                Price[i,DATE],S, Price[i,OPEN] );
        M := 0
    elif S>0 and Price[i,OPEN] < (1-p/100)*Price[i-1,CLOSE] then
```

```
        M := S * Price[i,OPEN];
        printf( 'on %d sell %.3f at %.3f, M=%.2f\n',
                Price[i,DATE],S, Price[i,OPEN], M );
        S := 0
    fi;
od;
# If there are any shares left, value them at the closing price
# of the last day
M := M + S*Price[length(Price),CLOSE]:
printf( 'final capital: $ %.2f\n', M );
```

```
on 20050119 buy 26.261 at 38.080
on 20050120 sell 26.261 at 35.390, M=929.36
on 20050202 buy 25.801 at 36.020
on 20050224 sell 25.801 at 30.430, M=785.13
on 20050324 buy 24.581 at 31.940
on 20050720 sell 24.581 at 34.210, M=840.93
on 20051019 buy 24.290 at 34.620
on 20051121 sell 24.290 at 41.260, M=1002.22
final capital: $ 1002.22
```

We see that this strategy does better than the previous one: we have not gained much, but at least we have not lost either.

These two examples show the essentials of simulation:

- the choice of model (strategy) with parameters,
- simulating (using this model) over a period of time,
- computing one or many results which will allow us to select the best model and the best parameter(s) for this model.

Notice that the programs have parameters, in this case p. Choosing the parameters so that the training data (i.e. the historical data) give best results is called training the model. This corresponds to "Simulate to obtain optimal parameters" in Figure 7.5. How this is done is the subject of the next section.

7.4.2 Optimization of simulation parameters

The historical sequence (training data) can be used to compute the optimal parameter values. We can view the running of the simulation as evaluation of a function whose arguments are the parameters and whose result is the final gain from the transactions. As a function of the parameters, simulations tend to be piecewise constant with

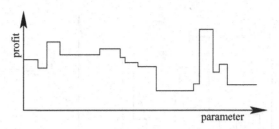

Figure 7.6   Example of a piecewise constant gain function with one parameter.

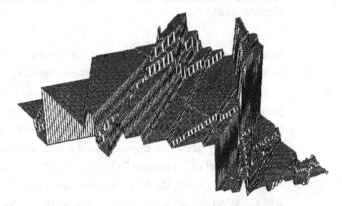

Figure 7.7   Example of a piecewise constant gain function with two parameters.

discontinuities. So, in one dimension (a model with only one parameter) the function to optimize could look like the example shown in Figure 7.6. In two dimensions the function could look like that shown in Figure 7.7.

This is a natural consequence of the parameters affecting the decisions in the computation. It is the typical situation when simulating decision processes. The peculiar shape of these functions is easy to explain. Let us first consider the case of a single parameter. If we change the parameter very little, and no decision is altered (a likely event if the change is very small), then the functional (the total gain/loss in this simulation period) will not change. That is, it is constant for some interval. As soon as we change the parameter in such a way that one or many decisions are different, then the resulting gain/loss will be different and we will have a jump/discontinuity.[4]

The situation for two parameters is analogous, except that the discontinuities will happen at linear (or other) combinations of the two parameters. A linear combination will imply a straight line. We see in Figure 7.7 the various surfaces limited by straight

---

[4] When p affects the prices, the function between discontinuities may not be constant, but a smooth function of p. Notice, however, that prices are actually rounded (to cents in the US markets) and hence the returns are always piecewise constant.

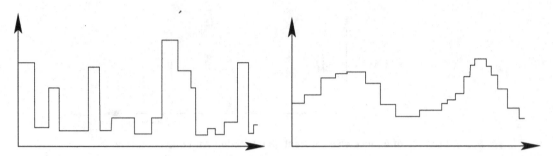

Figure 7.8   Poor model (too erratic, sharp maxima) versus good model (smoother shape, large maxima area).

lines. Optimization (maximization of the result of the simulation) is difficult, as we cannot rely on continuity, and even less on continuity of the derivatives.

Although very discontinuous, these functions should show some shape if they are good models for prediction. That is, if a small change in one parameter moves the results from the best possible to the worst possible, the model is not reliable and hence not very good. We expect a good model to show general areas of high values and some level of "continuity" (Figure 7.8). The more "continuous" the function is around its maximum, the more we can expect that minor changes in the parameters will not affect the optimum; hence we have greater reliability.

Finding the best parameters is a problem of maximization. We will study two methods for maximization of these piecewise-constant functions: `Analytic` and `PiecewiseConstMax`. All other methods are some form of brute force searching.

See the interactive exercise "Prediction."

The one-dimensional maximization algorithm `Analytic`

The analytic method finds all the points where the function has a discontinuity (or could have a discontinuity). Next it evaluates the function (runs the simulation) for the parameter just before and just after each discontinuity. The best, maximum, value is selected and the midpoint of the interval that produces it is returned. Let us analyze the steps of this algorithm in more detail.

The sources of discontinuities for the result of a simulation are conditional expressions, typically the expressions inside `if` statements. For example, in our first simulation we had

$$M > 0 \quad \text{and} \quad Open_i - p \geq Low_i + gap.$$

Assuming that our parameter is p, the first expression ($M > 0$) does not depend on p and will not cause any discontinuity in p. The second depends on p and the discontinuities will be on all points p such that

$$p = Open_i - Low_i - gap.$$

The second if statement will generate discontinuities at points

$$p = High_i - Open_i - gap.$$

If there are n periods of simulation we will generate $2n$ candidate values for p which may cause discontinuities. We normally sort these values to eliminate duplicates. The actual number of discontinuities is usually much less than $2n$, for the following reasons.

(i) The if expressions depend on other conditions ($M > 0$, $S > 0$) which could be false, making the test of the expression depending on p irrelevant, hence no decision will be changed and no discontinuity arises.
(ii) Some values may coincide (a point where two decisions will change), then a single point will be responsible for more than one change.
(iii) Some if statements may not be executed at all in certain cases, for example the second if statement in the programs above will not be executed if the first one is successful. These points will not cause discontinuities.
(iv) Finally some values may be out of the valid range for our parameter p and hence could never trigger a discontinuity, for example $p < 0$ in our example.

In the second phase we analyze all the p candidates in ascending order. To do this, we consider all the intervals defined by the sorted p values. (We include the first interval and the last, i.e. the interval lower than all p values and the interval higher than all p values.) For each interval we compute the result (run the simulation) for p set to the midpoint of the interval. The results of these simulations are the only possible values for the outcome for the case where the parameters are only used for decisions. From this process we retain the interval which gives the highest function value (or collection of intervals in the case that the function does not change across some points). Finally we return the midpoint of this interval.

The analytic method is very effective and economical. However, it requires knowledge of the program which does the simulation and requires a symbolic solution of the expressions in terms of the parameters. This can be done for some programs but not in general.

The next method works without knowledge of the program and will normally be called a "black box" approach.

The one-dimensional maximization algorithm `PiecewiseConstMax`

Assume we have a univariate function which is piecewise constant. This may be the result of a single parameter decision problem or the result of following a particular direction in a multivariate (multiparameter) decision problem. The main point is that $f(x)$ is piecewise constant,[5] as in Figure 7.6. We define a "plateau" to be a continuous interval, $x \in [a, b]$ where $f(x)$ has the same value. Our algorithm will find all the plateaux in a given range. Then by computing $f(x)$ for each plateau we can find the maximum. It is assumed that the number of different plateaux is not too large, or at least we are prepared to compute $f(x)$ for each different plateau. As seen for the analytic approach, the number of plateaux is linear in the number of data points and conditional expressions.

We will assume that different (disconnected) plateaux do not have the same $f(x)$ value. If this is not guaranteed, the test can be enhanced by a path signature as described later. With this property it is easy to identify when we have computed two points x and y on the same plateau. For such points, $f(x) = f(y)$.

Our procedure `PiecewiseConstMax` does a binary search between points which are not on the same plateau.

```
PiecewiseConstMax(x,fx,y,fy)
    if |x-y| < tolerance then return(max(fx,fy))
    elif fx = fy then return(fx)
    else m := (x+y)/2
        fm := f(m);
        return(max( PiecewiseConstMax(x,fx,m,fm),
                    PiecewiseConstMax(m,fm,y,fy) ))
    fi
end:
```

The `tolerance` parameter is either the machine epsilon or some sufficiently small value whose effect is judged to be insignificant for the purposes of the model. The above function is simple and elegant but returns only the value of the maximum. The next function sets global variables with the value of the maximum and the range of parameters for which it occurs.

```
global low_x, hi_x, max_f;
max_f := -infinity;
PiecewiseConstMax(x,f(x),y,f(y));
```

---

[5] If the simulation is realistic, then prices have to be rounded (e.g. to cents) and shares have to be bought or sold in units or even bigger lots like multiples of 100. This is the reason why piecewise-continuous functions are to be expected.

With the new definition:

```
PiecewiseConstMax(x,fx,y,fy)
   if y < x then PiecewiseConstMax(y,fy,x,fx)
   elif fx = fy then
        if fx > max_f then
             max_f := fx;
             low_x := x;
             hi_x := y;
        elif fx = max_f then
             if x < low_x then low_x := x fi;
             if y > hi_x then hi_x := y fi;
        fi;
   elif y-x < tolerance then
        if fx > fy then
             PiecewiseConstMax(x,fx,x,fx)
        else PiecewiseConstMax(y,fy,y,fy)
        fi
   else m := (x+y)/2;
        fm := f(m);
        PiecewiseConstMax(x,fx,m,fm);
        PiecewiseConstMax(m,fm,y,fy);
   fi
end;
```

The function is called with the two end points x, y and their function values. The complexity of this function is linear in the number of plateaux and the logarithm of the precision. It is possible to extend this function to higher dimensions but it will not be nearly as efficient, and its coding becomes too complicated. So we will consider this to be a univariate function only.

Path signature

We would like to assume that when $f(x) = f(y)$ then the parameter values x and y are on the same plateau. This means that the computation path for the parameter value x is the same as for the parameter value y. In normal circumstances, once we have different computation paths and different truncation errors (in the operations involved), this will always be the case. If we have a very simple model with little chance of random noise or if we want an extra assurance, we can compute a *path signature* which can be computed, at very little cost, to determine whether the computation path is the same or not.

A path signature is the sum of random values which depend on the branches taken in the computation and the period data. We give an example of the code with such a computation:

```
PathSig := 0;
for i from 1 to length(Price) do
    if M>0 and ...
        PathSig := PathSig + Sig1[i];
        S := M ...
            ⋮
    elif S>0 and ...
        PathSig := PathSig + Sig2[i];
            ⋮
```

The vectors `Sig1`, `Sig2`, etc. are vectors of random values, usually uniformly distributed in $U(-1, 1)$. For each $f(x)$ we will now also have the *Signature*(x) meaning the `PathSig` value at the end of the simulation. If we have $f(x) = f(y)$ and *Signature*$(x) =$ *Signature*(y) we can safely conclude that the parameter values x and y followed the same computation path and hence are on the same plateau.

Multidimensional methods

The methods in this section are usable for most optimization problems over complicated domains and non-smooth functions. They are very general, not just restricted to stock market prediction or model training.

Random directions The method of random directions is very general and useful every time that we have an efficient one-dimensional function. (See also Brent's algorithm used for multidimensional minimization, Appendix A, Sections A1.2.2 and A1.3.3.) In all cases $f(x)$ denotes the functional or gain/loss function evaluated for the parameter value x. The algorithm has the following structure:

 (i) choose a random starting point,
 (ii) choose a random direction,
 (iii) do `PiecewiseConstMax` in this direction,
 (iv) if a better point is found, replace the old point,
 (v) iterate until a large number of directions (e.g. 100) yields no improvements.

`f(h)` is the function which computes the simulation based on the global variables `x0` and `d`, namely it computes the function for the point `x0+h·d`. It is assumed that the range of h is between −1 and 1 (i.e. d will be of the right magnitude so that the space is properly covered).

x0, f0 will finally be the resulting pair of best parameter and value.

```
x0 := initial_point;      # should be random
f0 := f(0);               # computes function on x0
lastimprov := 0;
for iter to infinity do
    d := random_direction();
    max_f := -infinity;   # PiecewiseConstMax sets max_f
    PiecewiseConstMax(-1, f(-1), 0, f0 );
    PiecewiseConstMax( 0, f0, 1, f(1) );
    if max_f > f0 then    # direction was useful, new maximum found.
        x0 := x0 + d*(low_x + hi_x)/2
        f0 := max_f;
        lastimprov := iter
    elif iter - lastimprov > 100 then break
    fi
od;
```

Normally f(x) is defined as

```
f := proc(h)
    global x0, d;
    RunSimulation( x0+d*h)
end;
```

Spiral search This is a method suitable for two dimensions (two parameters) only. It can be easily visualized as following a shrinking spiral around a good point. The spiral should shrink relatively slowly, for example by a factor of 0.99 per complete turn. The algorithm has the following structure:

(i) initialize (random point and spiral radius),
(ii) follow the spiral inwards; an exponential spiral follows the equation $r = e^{\alpha\theta + i\beta\theta}$ in polar coordinates; we discretize and select some points over this inward spiral (the spiral is inwards if $\alpha < 0$ for increasing θ),
(iii) if a better point is found, replace it and increase the initial spiral radius by a factor of 2,
(iv) stop when the spiral is so small that it is completely contained in the plateau of the best point.

```
x0,y0 := initial_point    # (two dimensional)
f0 := f(x0,y0);
```

```
    r := initial_radius;        # (should not be too small)
    lastimprov := 0;
    h := 0.8683148;
    for iter to infinity do
        x1 := x0 + r*cos( iter*h );
        y1 := y0 + r*sin( iter*h );
        f1 := f(x1,y1);
        if f1 > f0 then
            f0 := f1;
            x0 := x1;
            y0 := y1;
            r  := 2*r;
            lastimprov := iter;
        elif iter - lastimprov > 100 then break
        fi
        r := r*0.998612
    od;
```

The peculiar constant h is designed to explore the two-dimensional space as evenly as possible. For the curious, $h = \pi/(3 + \phi)$, and $\phi = (\sqrt{5} - 1)/2 \approx 0.618$ is the ratio of the golden section. (See Appendix A, Section A1.2.2 for more details and another application of the golden section.) The factor 0.998612 multiplying r at the end of the loop is to make r shrink by 1% every *complete* turn, or by $0.99^{1/(6+2\phi)} \approx 0.998612$ in every step, since we check an average of $6 + 2\phi \approx 7.2$ points in each complete turn.

Shrinking hypercube This is a good algorithm for high dimensions (high number of parameters). It selects random points inside a hypercube around the best point. The hypercube is gradually shrunk. If a better point is found, it is selected and the size of the hypercube is doubled otherwise it is slightly reduced. This procedure is essentially similar to the spiral search:

  (i) initialize (initial random point and hypercube size),
 (ii) select a random point inside the hypercube,
(iii) if the point is better, replace the old point with the new one and double the hypercube edge length,
 (iv) otherwise shrink the hypercube edge length by 0.1%,
  (v) stop when a long sequence of points falls on the same plateau (same value of functional), choose this value as the best value.

```
x0 := initial_point;        # vector in  k dimensions
f0 := f(x0);
lastimprov := 0;
```

```
      r := 1;                    # initial width of hypercube,
                                 # should not be too small
      for iter to infinity do    # select a new point x1
          for i to n do
              x1[i] := x0[i] + r*U(-1,1);
                                 # U(-1,1) is a uniformly distributed
                                 # random number between -1 and 1
          od;
          f1 := f(x1);
          if f1 > f0 then
              x0 := x1;
              f0 := f1;
              lastimprov := iter;
              r := 2*r;
          elif iter - lastimprov > 100 then break
          fi;
          r := r*0.999
      od;
```

Adaptivity of spiral search and shrinking hypercube

These are adaptive algorithms whose parameters change dynamically to adapt to the particular problem being solved. The reason for doubling the size of the search space when a new maximum is found is to allow the algorithm to start off-scale and still be able to move arbitrarily far from the initial point. The reason for decreasing the length on each unsuccessful probe is to focus the search so that we can terminate. Normally we expect to compute many function evaluations between new maxima. If we expect to compute k function evaluations per new maximum, then the shrinking factor should be about $(\frac{1}{2})^{1/k}$. In average situations, the total amount of work is proportional to $\mathcal{O}(k \log m)$, where m is the total size of the search space (total number of plateaux).

Brute force search/grid search

The simplest way to find a maximum is by brute force. We can divide the space into a k-dimensional grid, compute one value for each grid point and choose the best point. This is normally very inefficient even for small values of k. Its only real justification is to use it in conjunction with the previous algorithms. For each point of the grid we start a spiral or shrinking hypercube search.

Notice that all the algorithms in more than one dimension are not guaranteed to succeed. They explore a part of the parameter space and are likely to return a local optimum of the function. In this sense they are very comparable to the early abort algorithm described for best basis in Section 4.3.1. A better picture of the quality of the optimization will be obtained by running these algorithms several times with different, random, starting points.

7.5 Confidence analysis

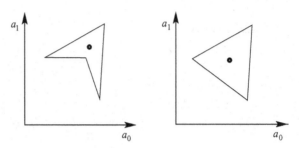

Figure 7.9   Good cases (the chosen point is as far as possible from the edges).

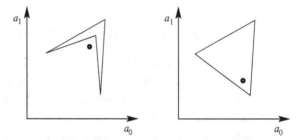

Figure 7.10   Bad cases (the chosen point is outside the region or near to the edges).

Once we find a local optimum there are three interesting values of the optimal region, which allow us to do a primary confidence analysis. Please notice that these tests do not replace the validation phase to be done with the test data (as described in the general diagrams, e.g. Figure 5.1). These tests are to be considered additional evidence/support, or in the negative case, sufficient reason to discard the model.

1. Center of gravity The first test is the center of gravity of the optimal region. (Remember: the function is piecewise constant, hence the maximum is an area not a single point.) A large region guarantees that small perturbations of the parameters have the best possible chance of remaining inside the optimal region. For this to be correct, the center of gravity should coincide with the center of the maximum region, see Figures 7.9 and 7.10. If the solution regions are convex, this is guaranteed. Models which are linear in the parameters split the solution space by hyperplanes and will give convex regions. The reason for choosing as solution the center of gravity is to choose a solution which is as far away from the borders as possible. Section 7.5.1 explains why this is desirable.

2. Volume of the optimal solution The second value of interest is the volume of the optimal solution (the area of the optimal solution in two dimensions, the hypervolume in n dimensions). The larger the volume, the more we can perturb the parameters and expect them to remain inside the optimal region. This is a direct measure of the reliability of our solution. Since absolute measures of volume are difficult to comprehend, this measure is most useful when comparing two solutions of the same problem.

3. Smoothness of the optimal area By smoothness we mean that the model shows the good feature of having some level of continuity as opposed to the bad feature of having large random variations. The smoothness is difficult to measure, hence we will use an alternative idea, which is to explore the areas neighboring the optimum. If the areas around the optimum have average or less than average gain, then this will be interpreted as lack of smoothness.

7.5.1 The notion of "transfer of error"

In reality, we are not interested in perturbing the parameters, we will never do that. However, the parameters are normally directly tied with the data, for example multiplied as in $a_0 \cdot High_{i-1}$. Hence a small perturbation in the data has the same effect as a small perturbation in the parameter, or we can say that the errors in the data can be "transferred" to errors in the parameters.

So a large volume of the optimal solution and choosing the center of gravity allows us to expect that small perturbations in the data will not change the optimal solution. In other words, a large volume and using the center of gravity of the solution will increase our confidence in the model.

7.5.2 Monte Carlo estimation

The volume, the center of gravity and the smoothness can all be estimated using a Monte Carlo technique. For this we bound the solution region by a small enclosing hypercube. Then we compute the function at random points inside the hypercube until enough hits (finding the optimum) are collected. "Enough" is a very ambiguous

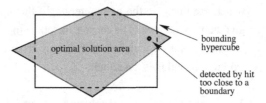

Figure 7.11   **The bounding hypercube is too small.**

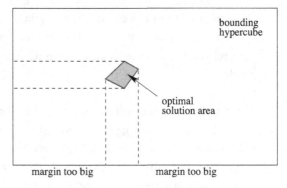

Figure 7.12   **The bounding hypercube is too big.**

term, in this case it means 20. ;-) The volume is then estimated by the ratio of hits to the total number of calculated points times the volume of the enclosing hypercube. The center of gravity is estimated by the average coordinates of all hits.

Exercise Prove that the relative error made by estimating the volume in this way is at least 19 times out of 20 less than $1.96/\sqrt{20}$.

The enclosing hypercube has to be computed adaptively. Recall that we do not know the shape of the solution nor its exact boundaries. The goal is to have a hypercube which leaves enough margin to insure that the solution is completely contained in the enclosing hypercube.

Let x_i be the coordinates of the hits (points which have a maximal functional). Let x_{lo}, x_{hi} be the coordinates of the enclosing hypercube. Our goal is to maintain for every dimension (i.e. parameter) the condition $x_{lo} + g = \min(x_i) \leq \max(x_i) = x_{hi} - g$, where g is a safety margin and is usually a small fraction of $\max(x_i) - \min(x_i)$ for this dimension, e.g. $g = \frac{1}{4}(\max(x_i) - \min(x_i))$.

If the margin becomes too small (Figure 7.11), we usually have to enlarge the hypercube in the desired dimensions and directions. Unless extreme care is taken, the Monte Carlo estimation should be restarted.

If the safety margin is too large (Figure 7.12), our method will be very inefficient (too few hits). Hence it is desirable to reduce the size of the hypercube to fit the solution area as tightly as possible. We always have at least one hit since we know the optimal value, and we use all the hits that we have to reduce the size of the hypercube, computing new boundaries that satisfy the relations stated above. When the hypercube is of the right size, we can estimate the smoothness by computing the average gain of all points inside the hypercube but outside the optimal solution area.

Let x_{lo} and x_{hi} be the lower and upper (vector) values of the bounding hypercube. Let $d = x_{hi} - x_{lo}$ be the vector of side lengths of the hypercube. Let n be the number of parameters, i.e. the dimension of x_{lo}, x_{hi} and x_i. Finally let us say that we performed $m > 20$ random samples of the hypercube and 20 hit the optimal solution (we assume that $x_1, x_2, \ldots, x_{20}$ are optimal, $x_{21}, \ldots, x_m$ are not):

$$\text{center of gravity} = \frac{1}{20} \sum_{i=1}^{20} x_i$$

$$\text{volume of solution} = \frac{20}{m} \prod_{j=1}^{n} d_j$$

$$\text{smoothness} = \frac{1}{f(x_1)(m-20)} \sum_{i=21}^{m} f(x_i).$$

The smoothness is always less than 1. The closer to 1, the better behaved the prediction, i.e. the smoother is the maximum area.

7.6 Dynamic programming

R. Bellman began the systematic study of dynamic programming in 1955. The word "programming," both here and in linear programming, refers to the use of a tabular solution method for optimization problems. Although optimization techniques incorporating elements of dynamic programming were known earlier, Bellman provided the subject with a solid mathematical basis. Dynamic programming typically applies to optimization problems in which a set of choices must be made in order to arrive at an optimal solution. As choices are made, subproblems of the same form often arise. Dynamic programming is effective when a given subproblem may arise from more than one partial set of choices; the key technique is to store, or "memorize," the solution to each such subproblem in case it should reappear. Without this memorizing, the number of possible choices/computations is exponential and typically intractable.

Each dynamic programming algorithm consists of three steps:

(i) define the function to be optimized recursively (usually in terms of optimal subproblems);

(ii) do a forward computation (bottom-up) to find the value of the optimal solution (avoiding recomputation of subproblems);

(iii) from this optimal value do a backward computation to find an optimal solution path (a set of choices which led to the optimal value).

The *forward computation* in step (ii) is basic for every dynamic programming problem. When the optimal solution path is of interest, it is often useful to store additional information during the forward computations of step (ii) to facilitate the construction of the optimal solution in step (iii). If however only the *value* of the optimal solution is of interest, the *traceback* in step (iii) can be omitted completely.

The key characteristics that an optimization problem must exhibit for dynamic programming to be applicable are related to the first step.

- An optimal solution is based on some optimal subproblems, i.e. the definition of the optimal value is recursive.
- The number of subproblems is not too large, which in practical terms means that the number of subproblems is polynomial in the size of the problem.
- The computation of one optimal value, based on the optimal subproblems (recursive definition), is not too expensive (i.e. polynomial time is required).

7.6.1 The idea of dynamic programming in more detail

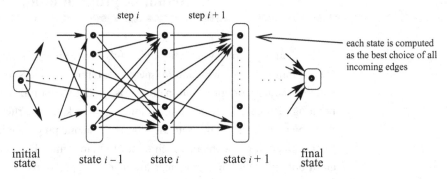

Figure 7.13   Schematic diagram of dynamic programming.

Each step represents one (or a family of) subproblems. For each subproblem we store its optimal value. The steps are organized in such a way that when computing a problem at step i all the subproblems needed are in steps before step i (Figure 7.13).

- Sequentially, for each step, we compute all the optimal values.
- Once the final state is reached, backtrack to find the path(s), which led to the optimal final state.

- The set of all transitions must not have cycles. It is also assumed that the total number of states is polynomial in the size of the problem and that the computation of each node will require polynomial time.
- Each transition is (possibly) associated with a cost. The computation of each state requires the use of optimal values of previous states and the cost of the transitions.
- The groups of states called *steps* are just used to organize the computation in a sequential way, they are not strictly needed.

7.7 Examples of dynamic programming

We shall further illustrate the principles of dynamic programming with three examples, the construction of optimal binary search trees (OBST), finding the optimal order of matrix multiplications and string alignment.

7.7.1 Optimal binary search trees

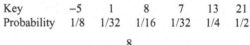

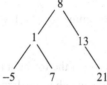

Figure 7.14   Example of a non-optimal BST (external nodes not shown).

A binary search tree (BST) is a tree where keys are stored in the internal nodes, the external nodes (leaves) are null nodes, and the keys are ordered lexicographically, i.e. for each internal node all the keys in the left subtree are less than the key in the node, and all the keys in the right subtree are greater.

When we know the probabilities of searching each one of the keys, it is quite easy to compute the expected cost of accessing the tree. An OBST is a BST which has minimal expected cost. An example is shown in Figure 7.14. The cost of searching a key is defined by the length of the path from the root to the key, for example the cost of finding "7" is 3. The expected cost of the entire tree is the sum of the costs of each key times the possibility of it being searched. So the expected cost of a search in this BST is:

$$E[cost] = 1 \cdot 1/16 + 2 \cdot (1/32 + 1/4) + 3 \cdot (1/8 + 1/32 + 1/2).$$

It is clear that this tree is not optimal, if the 21 is closer to the root, given its high probability, the tree will have a lower expected cost.

Criterion for an optimal tree Each optimal binary search tree is composed of a root (one of the keys) and (at most) two subtrees, the left and the right. The keys in the subtrees are uniquely defined once the root is defined (those smaller and those larger). These two subtrees must be optimal themselves.

Construction method The criterion for optimality gives a dynamic programming algorithm. For the root (and each node in turn) we select one value to be stored in the node. We have n possibilities to do that.

Once this choice is made, the sets of keys which go into the left subtree and right subtree are completely defined, because the tree is lexicographically ordered. The left and right subtrees are now constructed recursively (optimally). This gives the recursive definition of the optimal cost. Let p_i denote the probability of accessing key i, let $p_{i,j}$ denote the sum of the probabilities from p_i to p_j and let T_{ij} be the cost of an optimal tree constructed with the keys key_i, key_{i+1}, ..., key_j.

$$T_{ij} = \min_{k=i,\ldots,j} \left(p_{i,k-1}(1 + T_{i,k-1}) + p_k \cdot 1 + p_{k+1,j}(1 + T_{k+1,j}) \right) \frac{1}{p_{i,j}}.$$

The explanation of the formula is easy once we see that the first term corresponds to the left subtree, which is one level lower than the root, the second term corresponds to the root and the third term corresponds to the right subtree. Every cost is multiplied by its probability. We define $p_{i,i-1} = 0$ and $p_{i+1,i} = 0$, so T_{ii} simplifies to $T_{i,i} = p_i/p_{i,i} = 1$, which is the condition to terminate the recursion. This recursion requires exponential time to compute if applied directly. However, the optimal trees are only constructed over contiguous sets of keys, and there are at most $n(n+1)/2$ different sets of contiguous keys. This becomes evident below, when we construct the entries of the matrix $\mathbf{T}^*$: there are n sets of length one, $n-1$ contiguous sets of length 2, up to one set of length n, and $\sum_{i=1}^{n} i = n(n+1)/2$.

In this case we store the optimal cost of a subtree in a matrix $\mathbf{T}$. The matrix entry T_{ij} will contain the cost of an optimal subtree constructed with the keys i to j. We now fill the matrix diagonal by diagonal. It is customary to use a matrix $\mathbf{T}^*$ to save many multiplications and divisions. Let $T_{ij}^* = p_{ij} \cdot T_{ij}$ then

$$T_{ij}^* = p_{i,j} + \min_{k=i,\ldots,j} (T_{i,k-1}^* + T_{k+1,j}^*).$$

Note that for this formula to make sense, we have to define $T^*_{i,i-1} := 0$ (all entries below the main diagonal can be regarded as zero):

$$
\mathbf{T}^* =
\begin{array}{c|cccccc}
 & -5 & 1 & 7 & 8 & 13 & 21 \\
\hline
-5 & \frac{1}{8} & \frac{3}{16} & \frac{9}{32} & \frac{15}{32} & \frac{31}{32} & \frac{63}{32} \\
1 & & \frac{1}{32} & \frac{3}{32} & \frac{7}{32} & \frac{19}{32} & \frac{47}{32} \\
7 & & & \frac{1}{32} & \frac{1}{8} & \frac{15}{32} & \frac{21}{16} \\
8 & & & & \frac{1}{16} & \frac{3}{8} & \frac{19}{16} \\
13 & & & & & \frac{1}{4} & 1 \\
21 & & & & & & \frac{1}{2}
\end{array}
= \frac{1}{32} \times
\begin{pmatrix}
4 & 6 & 9 & 15 & 31 & 63 \\
 & 1 & 3 & 7 & 19 & 47 \\
 & & 1 & 4 & 15 & 42 \\
 & & & 2 & 12 & 38 \\
 & & & & 8 & 32 \\
 & & & & & 16
\end{pmatrix}.
$$

To understand the values in the matrix $\mathbf{T}^*$, let us reconstruct their calculation. Clearly $T^*_{ii} = p_{i,i} = p_i$, so in the main diagonal we have just the probabilities of each key. Now what is T^*_{34}?

$$
T^*_{34} = p_{3,4} + \min\left((T^*_{3,2} + T^*_{4,4}), (T^*_{3,3} + T^*_{5,4})\right),
$$

and using our definition $T^*_{i,i-1} := 0$ this simplifies to

$$
T^*_{34} = p_3 + p_4 + \min_{k=3,\dots,4}(p_3, p_4) = \frac{1}{32} + \frac{1}{16} + \frac{1}{32} = \frac{1}{8}.
$$

And as a last example:

$$
\begin{aligned}
T^*_{14} &= p_{1,4} + \min_{k=1,\dots,4}(T^*_{1,k-1} + T^*_{k+1,4}) \\
&= \sum_{i=1}^{4}(p_i) + \min(T^*_{1,0} + T^*_{2,4}, T^*_{1,1} + T^*_{3,4}, T^*_{1,2} + T^*_{4,4}, T^*_{1,3} + T^*_{5,4}) \\
&= \frac{1}{32} \cdot (4 + 1 + 1 + 2) + \min(0 + 7, 4 + 4, 6 + 2, 9 + 0) \\
&= \frac{1}{32} \cdot 15.
\end{aligned}
$$

The cost of the OBST is in $T_{1,n}$ ($T_{1,6}$ in our example). If we have n keys, we have to fill $n(n+1)/2$ values in the matrix. Each value costs $\mathcal{O}(n)$ operations to be computed, so the computation of the optimal cost is $\mathcal{O}(n^3)$ and not exponential. In this case the number of subproblems is $n(n+1)/2$.

Computing T^* is what we call the forward phase. It gives us the optimal cost, i.e. 63/32. To compute the tree that will give us this optimal cost we have to do the backtracking. That is, for each entry starting at T^*_{1n}, we find which $T^*_{1,k-1}$ and $T^*_{k+1,n}$ were used. This will determine the key at the root. Recursively we can reconstruct the whole OBST.

Interactive exercise "Optimal binary search tree"

What you can do

- Start the exercise by pressing the "Random keys" button or by clearing the tree and adding keys one by one starting from an empty tree. (Enter each key and its frequency in the two editable fields left to the buttons, then press "Add/ replace".)
- You can add nodes, delete nodes and change the frequency of a node. (The frequency is proportional to the probability of the node.) To choose the node, click on its index number, key number, or frequency in the table or enter its key value in the editable key field.
- When you click onto an entry in the optimal cost matrix **C**, the frequency matrix **F**, or the optimal root index matrix **R**, the corresponding subtree only is shown in the optimal binary search tree, and only the arcs corresponding to this optimal binary search tree are shown in the optimal root index matrix.

7.7.2 Optimal order of matrix multiplication

Given a product of matrices, for example **ABCDE**, there are many ways of evaluating this product (using the associativity of matrix multiplication). Let row and col denote the number of rows and columns of each matrix. For the above product to make sense we need to have $\mathrm{col}(\mathbf{A}) = \mathrm{row}(\mathbf{B})$, $\mathrm{col}(\mathbf{B}) = \mathrm{row}(\mathbf{C})$ etc.

The number of possible orderings for multiplying n matrices is very large. Let it be $N(n)$. Then by looking at the last product, which can be done in $n - 1$ different ways, we can define $N(n)$ recursively. For the above example, the product may be written as **A(BCDE)** or **(AB)(CDE)** or **(ABC)(DE)** or **(ABCD)E**:

$$N(n) := \sum_{i=1}^{n-1} N(i) \cdot N(n-i) \quad \text{for } n \geq 2, \tag{7.1}$$

with $\quad N(0) := 0 \quad$ and $N(1) := 1.$

We have $N(2) = N(1) \cdot N(1) = 1$, $N(3) = N(1) \cdot N(2) + N(2) \cdot N(1) = 2$, $N(4) = N(1) \cdot N(3) + N(2) \cdot N(2) + N(3) \cdot N(1) = 5$, etc.

This type of convolution sum is normally solved using generating functions. So we define the generating function to be

$$N(z) := \sum_{i=0}^{\infty} N(i)z^i = \sum_{i=1}^{\infty} N(i)z^i = z + \sum_{i=2}^{\infty} N(i)z^i. \tag{7.2}$$

From this follows:

$$[N(z)]^2 = \left(\sum_{j=1}^{\infty} N(j)z^j\right)\left(\sum_{k=1}^{\infty} N(k)z^k\right)$$

$$= \sum_{l=2}^{\infty} \sum_{j+k=l} N(j)z^j \cdot N(k)z^k$$

$$= \sum_{l=2}^{\infty} \sum_{j=1}^{l-1} N(j)z^j \cdot N(l-j)z^{l-j}. \tag{7.3}$$

But Equation (7.1), multiplied by z^n and this summed for $n = 1 \ldots \infty$ is:

$$\sum_{n=1}^{\infty} N(n) \cdot z^n = z + \sum_{n=2}^{\infty} \sum_{i=1}^{n-1} (N(i)z^i) \cdot (N(n-i)z^{n-i}) \tag{7.4}$$

$$= z + N(z)^2 \quad \text{(using (7.3)).} \tag{7.5}$$

$N(z) = z + [N(z)]^2$ is a quadratic equation in $N(z)$, yielding the two solutions:

$$N(z) = \frac{1}{2} \pm \frac{\sqrt{1-4z}}{2}.$$

Since $N(0) = 0$, we must choose the solution $N(z) = 1/2 - \sqrt{1-4z}/2$. You can use Maple to compute a series for this expression for $N(z)$, this still gives us only a few values.

The binomial expansion of $\sqrt{1+x}$ is

$$1 + \frac{1}{2}x - \frac{1 \cdot 1}{2 \cdot 4}x^2 + \frac{1 \cdot 1 \cdot 3}{2 \cdot 4 \cdot 6}x^3 - \frac{1 \cdot 1 \cdot 3 \cdot 5}{2 \cdot 4 \cdot 6 \cdot 8}x^4 + \cdots = \sum_{k=0}^{\infty} \binom{\frac{1}{2}}{k}x^k,$$

where

$$\binom{m}{k} := \frac{m(m-1)\cdots(m-k+1)}{1 \cdot 2 \cdot 3 \cdots k} \quad \text{and} \quad \binom{m}{0} = 1.$$

Applying this to $\sqrt{1-4z}$ the general term of the series in z becomes $\binom{1/2}{k}(-4z)^k$, and when choosing the negative sign in the solution of the quadratic equation, we get

$$N(z) = \frac{1}{2} - \frac{1}{2}\sum_{k=0}^{\infty} \binom{\frac{1}{2}}{k}(-4z)^k.$$

Comparing this expansion equation with Equation (7.4), we get:

$$\sum_{n=1}^{\infty} N(n) \cdot z^n = -\frac{1}{2} \cdot \sum_{k=1}^{\infty} \binom{\frac{1}{2}}{k}(-4z)^k. \tag{7.6}$$

Consequently:

$$N(n) = -\frac{1}{2}\binom{\frac{1}{2}}{n}(-4)^n$$

$$= -\frac{1}{2} \cdot \frac{\frac{1}{2}\left(-\frac{1}{2}\right)\left(-\frac{3}{2}\right)\left(-\frac{5}{2}\right)\cdots\left(-\frac{2n-3}{2}\right)}{1\cdot 2\cdot 3\cdots\cdot n}\cdot(-4)^n$$

$$= -\frac{1}{2} \cdot \frac{1(-1)(-3)(-5)\cdots(-2n+3)}{1\cdot 2\cdot 3\cdot 4\cdots n}\cdot(-2)^n$$

$$= -\frac{1}{2} \cdot \frac{1(-1)2(-3)4(-5)6\cdots(-2n+3)(2n-2)}{(1\cdot 2\cdot 3\cdot 4\cdots n)(2\cdot 4\cdot 6\cdots(2n-2))}\cdot(-2)^n$$

$$= \frac{(2n-2)!}{n!(n-1)!}$$

$$= \frac{1}{2n-1}\binom{2n-1}{n},$$

which is exponential in n. These numbers appear in many problems and are called Catalan numbers.

Generating functions are an important mathematical tool for solving this type of problem. In general they are useful when the recursion, in this case (7.1), can be transformed to a simple equation, namely (7.5), which can be solved explicitly. The procedure can be visualized schematically like this:

$$N(n) = \sum_{i=1}^{n-1} N(i)N(n-i) \quad\cdots\cdots\cdots\rightarrow\quad \frac{1}{2n-1}\binom{2n-1}{n}$$

generating function   invert generating function

$$N(z) = z + N(z)^2 \quad\longrightarrow\quad N(z) = \frac{1}{2} - \frac{\sqrt{1-4z}}{2}$$

solving

The results for $N(n)$ are shown in Table 7.1. This makes it clear that computing all possible orderings is out of the question in general.

The cost of calculating $A \cdot B$ with dimensions $\text{row}(A) \times \text{col}(A)$ and $\text{row}(B) \times \text{col}(B)$ and $\text{col}(A) = \text{row}(B)$ is: $\text{row}(A) \cdot \text{col}(A) \cdot \text{col}(B)$.

So, if we define T_{ij} as the cost of multiplying matrices i up to j, the optimal cost is

$$T_{ij} = \min_{k=i,\ldots,j-1} \left(T_{ik} + T_{k+1,j} + \text{row}(i) \cdot \text{col}(k) \cdot \text{col}(j)\right).$$

The cost of T_{ii} is 0. Finding the optimal solution is now completely analogous to finding an optimal binary search tree.

Exercises Five matrices A, B, C, D, E are given, of the following dimensions: $A = 2 \times 13$; $B = 13 \times 4$; $C = 4 \times 15$; $D = 15 \times 6$; $E = 6 \times 7$.

n	$N(n)$
Table 7.1 The number of different matrix multiplications for n matrices	
2	1
3	2
4	5
5	14
6	42
7	132
8	429
9	1430
10	4862
15	2674440
20	1767263190
25	1289904147324
30	1002242216651368

1. Find the best order of multiplication!
2. Find the cost of the multiplication!

7.7.3 String matching

String matching tries to establish a correspondence between two strings to show their similarities. The question posed is, how can we align two strings (one on top of the other) with minimal penalty? As penalty we define a very simple measure (also called Levenshtein's distance):

- deletion (or insertion in the other string) is penalized by 1 unit;
- mismatch is penalized by 1 unit.

Example

```
A  B  C  D              A  —  B  C  D
              ⟶         |     |  |        ⟶  2 units
A  X  B  C              A  X  B  C  —
```

The way to derive a recursion to find this optimum is to study the last position of the alignment. In an optimal match there are only three possibilities for the last matched position:

- match or mismatch,
- one insert in the upper string (or a deletion in the lower string),
- one insert in the lower string (or a deletion in the upper string).

Excluding the last position, the alignment should be optimal too. So we have optimal subproblems and overlapping substructure again, leading again to dynamic programming as the method of choice to solve string matching problems.

Definition Let $X = (x_1, x_2, \ldots, x_m)$ and $Y = (y_1, y_2, \ldots, y_n)$ be sequences. Let $M_{m,n}$ denote the minimum penalty of aligning X and Y. In general let $M_{i,j}$ denote the minimum penalty for aligning $x_1, x_2, \ldots, x_i$ against $y_1, y_2, \ldots, y_i$. The above three possibilities are translated into

$$M_{m,n} = \min(M_{m-1,n-1} + \delta_{mn}, M_{m,n-1} + 1, M_{m-1,n} + 1)$$

$$\text{where } \delta_{m,n} = \begin{cases} 1 & \text{for} \quad x_m \neq y_n \\ 0 & \text{for} \quad x_m = y_n. \end{cases}$$

As border conditions we use $M_{0,0} = 0$ and all other values outside (i.e. matrix-elements with negative indices) are ∞. M can be computed row by row (top to bottom) or column by column (left to right). It is clear that computing $M_{m,n}$ requires $\mathcal{O}(mn)$ work. If we are interested in the optimal value alone, we only need to keep one column (or one row) as we do the computation. Hence the space required for the computation of the optimum penalty is $\mathcal{O}(\min(m, n))$.

Example

		B	I	O	L	O	G	I	S	C	H	E	M	E	D	I	Z	I	N
	0	1	2	3	4	5	6	7	8	9	10	11	12	13	14	15	16	17	18
B	1	0	1	2	3	4	5	6	7	8	9	10	11	12	13	14	15	16	17
I	2	1	0	1	2	3	4	5	6	7	8	9	10	11	12	13	14	15	16
O	3	2	1	0	1	2	3	4	5	6	7	8	9	10	11	12	13	14	15
L	4	3	2	1	0	1	2	3	4	5	6	7	8	9	10	11	12	13	14
O	5	4	3	2	1	0	1	2	3	4	5	6	7	8	9	10	11	12	13
G	6	5	4	3	2	1	0	1	2	3	4	5	6	7	8	9	10	11	12
I	7	6	5	4	3	2	1	0	1	2	3	4	5	6	7	8	9	10	11
C	8	7	6	5	4	3	2	1	1	1	2	3	4	5	6	7	8	9	10
A	9	8	7	6	5	4	3	2	2	2	2	3	4	5	6	7	8	9	10
L	10	9	8	7	6	5	4	3	3	3	3	3	4	5	6	7	8	9	10
	11	10	9	8	7	6	5	4	4	4	4	4	5	6	7	8	9	10	11
M	12	11	10	9	8	7	6	5	5	5	5	5	4	5	6	7	8	9	10
E	13	12	11	10	9	8	7	6	6	6	6	6	5	4	5	6	7	8	9
D	14	13	12	11	10	9	8	7	7	7	7	7	6	5	4	5	6	7	8
I	15	14	13	12	11	10	9	8	8	8	8	8	7	6	5	4	5	6	7
C	16	15	14	13	12	11	10	9	9	9	9	9	8	7	6	5	5	6	7
I	17	16	15	14	13	12	11	10	10	10	10	10	9	8	7	6	6	5	6
N	18	17	16	15	14	13	12	11	11	11	11	11	10	9	8	7	7	6	5
E	19	18	17	16	15	14	13	12	12	12	12	12	11	10	9	8	8	7	6

The optimal alignment is recovered from backtracking from the $M(m, n)$ position. Ambiguities in the backtracking are not important, they just mean that there is more than one possible alignment with optimal cost. An optimal alignment would be:

```
B I O L O G I S C H E - M E D I Z I N -
B I O L O G I - C A L   M E D I C I N E
```

Backtracking (recovering the optimal alignment using only $\mathcal{O}(\min(m, n))$ memory)

Computing the optimum requires only one row or column in memory, but recovering the alignment requires backtracking, which makes it necessary to store the whole matrix. This may be too expensive in terms of storage. It is possible to compute the alignment in essentially the same time, but using only storage proportional to one row or column.

First we observe that the problem is symmetric, i.e. we can align the sequences from left to right or from right to left. We use this observation to construct a divide and conquer procedure to find the optimal alignment.

Second we choose a row where we split the matrix in two halves, thus dividing the problem into two subproblems. (We will discuss the solution using rows only from now on.) This row should be closest to the middle of the matrix. In our example we use row 9, and we compute the matrix M up to this row. We have now:

		B	I	O	L	O	G	I	S	C	H	E	M	E	D	I	Z	I	N
	0	1	2	3	4	5	6	7	8	9	10	11	12	13	14	15	16	17	18
B	1	0	1	2	3	4	5	6	7	8	9	10	11	12	13	14	15	16	17
I	2	1	0	1	2	3	4	5	6	7	8	9	10	11	12	13	14	15	16
O	3	2	1	0	1	2	3	4	5	6	7	8	9	10	11	12	13	14	15
L	4	3	2	1	0	1	2	3	4	5	6	7	8	9	10	11	12	13	14
O	5	4	3	2	1	0	1	2	3	4	5	6	7	8	9	10	11	12	13
G	6	5	4	3	2	1	0	1	2	3	4	5	6	7	8	9	10	11	12
I	7	6	5	4	3	2	1	0	1	2	3	4	5	6	7	8	9	10	11
C	8	7	6	5	4	3	2	1	1	1	2	3	4	5	6	7	8	9	10
A	9	8	7	6	5	4	3	2	2	2	2	3	4	5	6	7	8	9	10

We only keep the last row of this matrix in memory.

Next we compute the bottom part of the matrix in reverse, i.e. we reverse both sequences and compute the matrix M. In our example this results in:

		N	I	Z	I	D	E	M	E	H	C	S	I	G	O	L	O	I	B
	0	1	2	3	4	5	6	7	8	9	10	11	12	13	14	15	16	17	18
E	1	1	2	3	4	5	5	6	7	8	9	10	11	12	13	14	15	16	17
N	2	1	2	3	4	5	6	6	7	8	9	10	11	12	13	14	15	16	17
I	3	2	1	2	3	4	5	6	7	8	9	10	10	11	12	13	14	15	16
C	4	3	2	2	3	4	5	6	7	8	8	9	10	11	12	13	14	15	16
I	5	4	3	3	2	3	4	5	6	7	8	9	9	10	11	12	13	14	15
D	6	5	4	4	3	2	3	4	5	6	7	8	9	10	11	12	13	14	15
E	7	6	5	5	4	3	2	3	4	5	6	7	8	9	10	11	12	13	14
M	8	7	6	6	5	4	3	2	3	4	5	6	7	8	9	10	11	12	13
	9	8	7	7	6	5	4	3	3	4	5	6	7	8	9	10	11	12	13
L	10	9	8	8	7	6	5	4	4	4	5	6	7	8	9	9	10	11	12

Again, we only keep the last row of this matrix in memory.

Now, with these two rows we can decide at which point the optimal penalty is achieved. The total penalty is going to be the sum of the penalties of the last row in the first matrix (row i) and the penalties in the last row of the second matrix, i.e. row $(m + 1 - i)$ in the original matrix. (In our example the last row in the second matrix corresponds to row 10 of the original matrix.)

Notice that we have to add entry i in the last row of the top matrix and entry $(m - i)$ in the last row of the bottom matrix. In this way we find an i for which $\text{top}_i + \text{bottom}_{m+1-i}$ is minimal.

In our example this happens for $i = 9$ (and $i = 8$ or $i = 7$), for which the penalty is 6, as we already calculated using the full matrix.

Now we know that backtracking will go through row 9, column 9 of the original matrix, and we can apply the same procedure recursively to the subsequences BIOLOGICA + BIOLOGISC and to L_MEDICINE + HEMEDIZIN.

The total cost is proportional to the number of entries in the matrix **M** that we have to compute. The first division requires mn entries. The two next computations require exactly $mn/2$ both together, if the chosen row splits the matrix in exact halves. So the total computation requires

$$mn + \frac{mn}{2} + \frac{mn}{4} + \cdots \leq 2mn \quad \text{or } \mathcal{O}(mn).$$

FURTHER READING

T. H. Cormen, C. E. Leiserson and R. L. Rivest, *Introduction to Algorithms*, MIT Press & McGraw-Hill, 1990.

Phylogenetic tree construction

8.1 Introduction

A phylogenetic tree is a construction which describes the ancestor–descendant relations of a set of entities. When the entities are, for example, languages, the tree describes how languages have evolved. Usually the leaves of the tree represent the present-day entities (no descendants yet) and the internal nodes represent common ancestors which existed in the past. Most phylogenetic trees have an implicit notion of evolution from ancestors to current-day entities.

Phylogenetic tree construction is an important area of research in biology with applications also in linguistics, archeology, literature, mythology, history, software genealogy, etc.

8.1.1 Applications

Two of the most prominent examples are the tree of life and historical linguistics.

The tree of life

Ancient origins A first classification scheme for animals was proposed by ARISTOTLE, in his *De Partibus Animalium*[1] around 350 BC. There he already distinguished and classified more than 500 different species in classes – with remarkable accuracy. Some of his insights were forgotten during the next 2000 years, for example that dolphins and whales are not fish since they have lungs and are viviparous.

A first classification of plants was given by Aristotle's disciple THEOPHRAS-TUS OF ERESUS in his work *De Historia Plantarum*[2] about 320 BC. He classified

---

[1] http://etext.virginia.edu/toc/modeng/public/AriPaan.html
[2] http://etext.virginia.edu/kinney/plant.html
www.abocamuseum.it/uk/bibliothecaantiqua/Book_View.asp?
Id_Book=161&Display=E

plants as trees, bushes, shrubs etc., basing his classification on the form of growth. This method is not correct from a modern viewpoint.

First generation (taxonomy) The first modern classification scheme was developed by CAROLUS LINNAEUS, the founder of modern taxonomy; in his *Systema Naturae*[3] (1735) he based his system on the sexuality of plants, i.e. the number of stamens in the flower. Linnaeus introduced the modern binomial nomenclature in biology, designating species with two names, such as *Homo sapiens*, the first referring to the genus (which normally contains multiple species) and the second designating the species itself. In the tenth edition of the *Systema Naturae* (1758) he extended this classification scheme to animals.

Second generation (evolution) JEAN-BAPTISTE DE LAMARCK was the first to publish the idea of a genesis of plants, animals and mankind, i.e. a gradual change and development in his *Philosophie zoologique*[4] (1809). He believed in the hereditary transmission of acquired traits, for example that giraffes stretching their necks to reach higher leaves bequeath their longer necks to their offspring. His idea of an evolution of species was in direct contrast to the beliefs of the scientific community of his time, most notably the very influential biologist and paleontologist GEORGES CUVIER, who believed (correctly) in the existence of catastrophic revolutions on Earth[5] (extinguishing some species) but postulating from this the permanence of the surviving species, a conclusion based (incorrectly) on his findings about the modern ibis and the ibis in old Egypt. Nevertheless Cuvier gave a quite good classification scheme of animals.[6]

CHARLES DARWIN introduced the notion of evolution, i.e. a gradual change by selection of the fittest in his famous book *The Origin of Species*[7] (1859). Using the giraffe example, Darwin would argue that giraffes come with some variation in neck length and during hard times the ones with longer necks find more food at the top of the trees and hence survive, producing in the long term more offspring with longer necks. Lacking better, objective criteria, he inferred the development and relationships of species from their phylogenetic similarities.

ERNST HAECKEL was one of the most eloquent proponents of Darwin's evolution theory. He mapped a genealogical tree relating all animal life in his *The History of Creation* (1868) and *The Evolution of Man*[8] (1874).

---

[3] www.deutsches-museum.de/bib/entdeckt/alt_buch/buch0699.htm

[4] www.ucl.ac.uk/taxome/jim/Mim/lamarck_contents.html

[5] http://home.tiscalinet.ch/biografien/sources/cuvier_revo.htm

[6] http://visualiseur.bnf.fr/Visualiseur?Destination=Gallica&O=NUMM-88103

[7] www.gutenberg.org/etext/1228

[8] www.gutenberg.org/etext/8700 or
http://etext.teamnesbitt.com/books/etext/etext04/vlmn110.txt.html
http://caliban.mpiz-koeln.mpg.de/~stueber/haeckel/natuerliche/index.html

Third generation (phylogenies) The growing knowledge led to phylogenies based on shared attributes leading, in the end, to character compatibility models and cladistics. This started with WILLI HENNIG's *Grundzüge einer Theorie der phylogenetischen Systematik* (1950), which built on WALTER ZIMMERMANN's *Die Phylogenie der Pflanzen* of 1930.

Fourth generation (molecular sequence data) When molecular sequence data became available, phylogenies based on these objective data could be developed. This started in the 1960s with LINUS PAULING and EMILE ZUCKERKANDL who proposed the existence of a molecular clock from their work on hemoglobin.[9] Their theory had to be modified, but the basic idea of a molecular clock (limited to selection-neutral mutations) survives in the work of MOTOO KIMURA.[10] The ideas were extended with the notion of evolutionary distances. These methods are the only ones which can reconstruct the relations between the three kingdoms of life (Archaea, Bacteria and Eucaryota). Research on the basis of molecular sequence data is continuing. Many contributions to this field are made by numerous people.

To illustrate the molecular based methods we will use a protein found in many species, triosephosphate isomerase. The protein sequences for lettuce, rice, human, rhesus monkey and mosquito can be seen in Section 8.1.4.

Historical linguistics

There are over 5000 different human languages in the world. One basic question is, how are they related? Historical linguistics is the branch of linguistics that focuses on the interconnections between different languages in the world and/or their historical development. Historical linguists investigate how languages evolved and changed through time.

Basically languages have evolved in the past from populations separating geographically and/or mixing with others and slowly changing their vocabulary, pronunciation and grammar. There are still about 125 completely independent families of languages. Indo-European is one of those families which groups most of the languages spoken in Europe and elsewhere. The **Indo-European tree of languages** has been carefully reconstructed and is a second example of a phylogenetic tree (see Figure 8.1).

[9] E. Zuckerkandl and L. Pauling, Molecules as documents of evolutionary history, *Journal of Theoretical Biology*, **8**(2) (1965) 357–366.
[10] M. Kimura, *The Neutral Theory of Molecular Evolution*, Cambridge University Press, 1983.

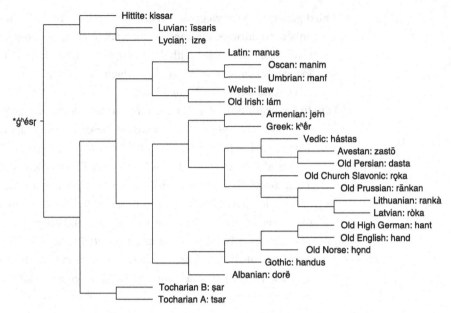

Figure 8.1   The Indo-European tree of languages for the word "hand."
· D. RINGE, T. WARNOW, and A. TAYLOR, Indo-European and Computational Cladistics,
Transactions of the Philological Society, **100**(1) (2002) 59ff.

BASIC

8.1.2 What it is about

The problem in phylogenetic tree construction, is to find the tree which best describes
the relationships between objects in a set. Usually and in our examples, the objects
are species.

Phylogenetic tree construction applied to biology is based on **cladistics**, the systematic classification of species based on shared characteristics thought to derive
from a common ancestor. The basic assumptions are as follows:

(i) Any group of organisms are related by descent from a common ancestor.
(ii) There is a bifurcation pattern of cladogenesis which, without loss of generality,
     can be assumed to be binary.
(iii) Changes in characteristics occur in lineages over time.

Under these conditions, phylogenetic trees can be represented by binary trees. The
leaves represent the objects which have no descendants, i.e. present-day species. The
internal nodes represent ancestors common to all descendants below. The root of
the tree represents the ancestor of all objects/species. The path from an ancestor to
a descendant is called a "lineage." Although other patterns of organization/evolution

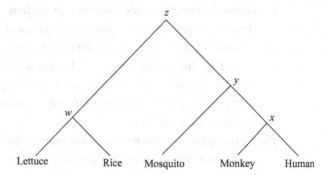

Figure 8.2 **Phylogenetic tree of five species.**

are possible, for example networks, we will restrict ourselves to trees which are by far the most common and useful structure.

We are normally confronted with a present-day set of species, for example lettuce, rice, mosquito, chimpanzee and human (see Figure 8.2) for which we want to reconstruct the evolutionary history based on their observable similarities and differences. The evolutionary branching process happened millions of years ago. The ancestors of chimpanzees and humans speciated some 10 million years ago (at point x in Figure 8.2). Plants and animals diverged more than a billion years ago (at point z in Figure 8.2).

To build a phylogenetic tree we need to have some measurable criteria which allow us to decide which tree fits the data best. Most phylogenetic trees associate a distance/length with each branch of the tree, the time or amount of change that happened in that part of the evolution, for example the branch $y - x$ in Figure 8.2 measures the time/change between the common ancestor of mosquitoes and humans and the common ancestor of chimps and humans.

Building a phylogenetic tree always implies finding the topology (shape) of the tree, and usually also finding the branch lengths. As it turns out, determining the topology is much more difficult than estimating the branch lengths.

As mentioned above, it is difficult or impossible to obtain information about the internal nodes x, y, z of the tree. We only know the leaves (lettuce, rice, mosquito, monkey, human) of the tree which represent present-day species. So we will restrict ourselves to this case.

BASIC

8.1.3 Classification types

There are three main types of trees, which can be distinguished by the classification method used to build them.

Phylogenetic trees based on distance information To construct a phylogenetic tree based on distance information, the definition of a distance between the species considered (i.e. the leaves of the tree) is required. Then a tree is constructed (topology and branch length) so that the distances between any pair of leaves can be mapped over the tree as accurately as possible. Mapping the distance between a pair of leaves over the tree means taking the sum of all branch lengths between these two leaves.

Phylogenetic trees based on character information A character or trait is a discrete property of a species (leaf). It can be assigned to each species, for example:

- mammal (all species are either mammals or not),
- multicellular (all species are either unicellular or multicellular),
- eye color (gray, blue, green, brown, none),
- in general, any property with a finite number of states where changing from one state to any other state is roughly equally likely.

Under this classification internal nodes are assigned characters to minimize the total number of character changes. The method that minimizes the number of character changes is called the most parsimonious or *parsimony* for short. The tree selected is the topology which has maximum parsimony.

We will look at this in detail in Section 8.5.

Phylogenetic trees based on probabilistic methods (maximum likelihood methods) This classification is based on the likelihood (probability) of a certain tree explaining a given set of species. The species are described by characteristics which evolve according to some probabilistic model. This is normally applied to genomic sequences (either DNA or amino acid sequences). The tree (topology and branch lengths) which maximizes the likelihood of evolving from a common root to all leaves is selected. We will not cover this subject in detail.

8.1.4 Phylogenies from protein sequences

This is a sample of peptide sequences of triosephosphate isomerase taken from five different species. The sequences are shown aligned to each other in what is called a *multiple sequence alignment*.

```
Lett_____IQVA
Rice _MGRKFFVGGNWKCNGTTDQVDKIVKILNEGQIASTDVVEVVVSPPYVFLPVVKSQLRPEIQVA
Chim APSRKFFVGGNWKMNGRKQNLGELIGTLNAAKVPAD_TEVVCAPPTAYIDFARQKLDPKIAVA
```

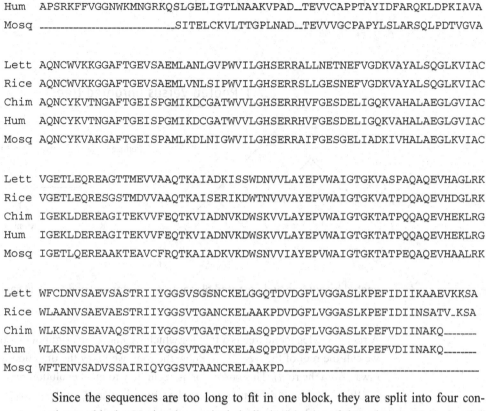

```
Hum   APSRKFFVGGNWKMNGRKQSLGELIGTLNAAKVPAD_TEVVCAPPTAYIDFARQKLDPKIAVA
Mosq  _____SITELCKVLTTGPLNAD_TEVVGCPAPYLSLARSQLPDTVGVA


Lett  AQNCWVKKGGAFTGEVSAEMLANLGVPWVILGHSERRALLNETNEFVGDKVAYALSQGLKVIAC
Rice  AQNCWVKKGGAFTGEVSAEMLVNLSIPWVILGHSERRSLLGESNEFVGDKVAYALSQGLKVIAC
Chim  AQNCYKVTNGAFTGEISPGMIKDCGATWVVLGHSERRHVFGESDELIGQKVAHALAEGLGVIAC
Hum   AQNCYKVTNGAFTGEISPGMIKDCGATWVVLGHSERRHVFGESDELIGQKVAHALAEGLGVIAC
Mosq  AQNCYKVAKGAFTGEISPAMLKDLNIGWVILGHSERRAIFGESGELIADKIVHALAEGLKVIAC


Lett  VGETLEQREAGTTMEVVAAQTKAIADKISSWDNVVLAYEPVWAIGTGKVASPAQAQEVHAGLRK
Rice  VGETLEQRESGSTMDVVAAQTKAISERIKDWTNVVVAYEPVWAIGTGKVATPDQAQEVHDGLRK
Chim  IGEKLDEREAGITEKVVFEQTKVIADNVKDWSKVVLAYEPVWAIGTGKTATPQQAQEVHEKLRG
Hum   IGEKLDEREAGITEKVVFEQTKVIADNVKDWSKVVLAYEPVWAIGTGKTATPQQAQEVHEKLRG
Mosq  IGETLQEREAAKTEAVCFRQTKAIADKVKDWSNVVIAYEPVWAIGTGKTATPEQAQEVHAALRK


Lett  WFCDNVSAEVSASTRIIYGGSVSGSNCKELGGQTDVDGFLVGGASLKPEFIDIIKAAEVKKSA
Rice  WLAANVSAEVAESTRIIYGGSVTGANCKELAAKPDVDGFLVGGASLKPEFIDIINSATV_KSA
Chim  WLKSNVSEAVAQSTRIIYGGSVTGATCKELASQPDVDGFLVGGASLKPEFVDIINAKQ_____
Hum   WLKSNVSDAVAQSTRIIYGGSVTGATCKELASQPDVDGFLVGGASLKPEFVDIINAKQ_____
Mosq  WFTENVSADVSSAIRIQYGGSVTAANCRELAAKPD_____
```

Since the sequences are too long to fit in one block, they are split into four contiguous blocks. Notice the vertical similarity/identity of the columns. Note also that the more distant species, e.g. human–lettuce, have many more differences than the "closer" species, e.g. human–monkey. This is the basis for inferring phylogenies from molecular sequence data.

The problem is defined as follows. Find a tree of the species which explains the differences between the sequences in the simplest or most probable way.

8.2 Tree building

The complexity of building a phylogenetic tree can be measured by the time needed to build a tree which satisfies the criteria. This normally means finding the topology and optimal branch lengths.

Except for some simple cases explained in Section 8.6, building phylogenetic trees is NP-complete and hence very hard. To solve real problems (sometimes with hundreds of leaves) we use approximation algorithms. Hence we have no guarantee of having found the optimal tree. This makes the problem of finding good trees fast very interesting and challenging.

8.2.1 Rooted versus unrooted trees

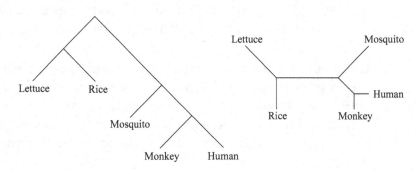

Rooted binary tree and unrooted planar tree with ternary nodes of the same species.

A rooted phylogenetic tree is a tree for which we know the origin of evolution which we call the root of the tree. If we know the root, we know the direction of evolution (away from the root). This corresponds accurately to most evolution processes.

Often we cannot determine where the origin is placed. In these cases (which are common) we build an unrooted tree which is normally represented in a planar way with all internal nodes being ternary nodes (Figure 8.3).

An unrooted tree with n leaves has $n - 2$ internal nodes and $2n - 3$ branches. Such a tree can be converted into a rooted tree if we know on which branch we should position the origin. This means that an unrooted tree can be converted to $2n - 3$ different rooted trees which describe the evolution of the same species (see Figure 8.4).

When we have an unrooted tree – and most of the important methods produce unrooted trees – we need extra information to place the origin. This is relatively easy to do if we can add to our set of species an *outgroup*, i.e. a species which can be related to our set, but which is known to be outside of our set. By building an unrooted tree with the outgroup we can easily identify the origin. It will be where the outgroup links to the rest of the tree.

For our example we could add a bacterium, say *E. coli*, and then obtain the tree shown in Figure 8.5, which shows that the third version of the rooted trees in (Figure 8.4) is the correct one.

The number of different topologies with n leaves grows very rapidly; an example is given in Table 8.1.

n	Unrooted binary	Rooted binary	Rooted n-ary
Table 8.1 Number of different trees with n leaves $n!! := n(n-2)(n-4)\ldots$ (until 2 or 1)			
2	1	1	1
3	1	3	4
4	3	15	27
5	15	105	256
6	105	945	3125
10	2027025	34459425	387420489
20	2.21×10^{20}	8.20×10^{21}	1.97×10^{24}
50	2.84×10^{74}	2.75×10^{76}	6.60×10^{82}
n	$(2n-5)!!$	$(2n-3)!!$	$(n-1)^{n-1}$

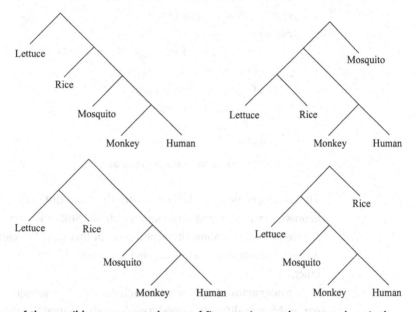

Figure 8.4   Four of the possible seven rooted trees of five species – only one topology is the correct one.

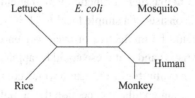

Figure 8.5   The unrooted tree with *E. coli* as outgroup.

Note that the approximate number of atoms in the visible universe is "only" about 10^{80}. These calculations are for the topology only. The set of trees on n vertices with real branch lengths is not even countable.

8.2.2 The constructor–heuristic–evaluator (CHE) optimization strategy

In general the tree building algorithms which we use follow the constructor–heuristic–evaluator (CHE) optimization strategy. This strategy is applicable to a large class of difficult optimization problems, and has the following common skeleton:

```
sol := Constructor(problem);
vsol:= Evaluator(sol);
repeat
    new := Heuristic(problem, sol);
    vnew:= Evaluator(new);
    if vnew > vsol then
        sol := new;
        vsol:= vnew
    fi
until termination_condition
```

This is a very general skeleton where the main difficulties are in the constructor, heuristic, evaluator and termination criteria. Still it is very useful to organize the solution of these optimization problems in this way. We can study a repertoire of different constructors/heuristics and evaluators which can be combined with each other.

A **constructor** is a function which constructs a solution to our problem. Since our problem is difficult, the solution is rarely optimal. Recall that we are solving an optimization problem and the difficulty is in finding an optimal solution, not just any solution.

A **heuristic** is a function which attempts to improve an existing solution. Usually this consists of a simple local fix. It is not guaranteed that the heuristic will succeed, it could be ineffective or even worsen our solution. It is clear that there are many heuristics and that these could be applied to different parts of the problem.

An **evaluator** is a function which computes the objective function to be optimized. We assume without exception that a higher value means a better solution.

The CHE strategy will be used to solve the hard tree problems, but there are three simple special cases which result in fast tree construction; see Section 8.6.

8.3 Trees based on distance information

To construct a tree of n leaves based on distance information, the following input is necessary

- an $n \times n$ symmetric matrix with positive distances,
- optionally an $n \times n$ symmetric matrix of the variances of the distances (a measure of the error that the distances may have).

The output is an unrooted tree with n leaves (present-day species) and $n - 2$ internal nodes (representing ancestral species) connected by $2n - 3$ branches, each one with its length. In graph theory terms this is an unrooted tree with each edge weighted by its length and each internal node of degree 3.

What we need are the following.

(i) A measure of the distance between species (and optionally their variance).
(ii) A method for constructing good initial topologies (constructor).
(iii) A method of fitting the distances between sequences to a tree topology. This needs branch lengths. The error of this approximation is what we want to optimize, hence computing this error is our evaluator.
(iv) A search strategy for better topologies (heuristics).

8.3.1 Measures of distance

To be able to determine the topology and branch lengths we need to estimate distances between species. The *distance* between species A and B, $d(A, B)$, should measure the length of the evolutionary path between X and A plus the length of the path between X and B (Figure 8.6).

As a first approximation we may think of the distance as the time elapsed since the divergence of the two species. This is ideal, but time is very difficult to measure when we do not have traces of the ancestors. So instead we measure distances in other ways. When we have molecular sequences we measure distance by the amount of change (mutations) between the sequences.

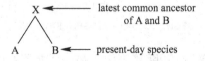

Figure 8.6   Two species with a common ancestor.

Distance functions must follow the classical distance properties:

$$d(A, A) = 0 \qquad \text{identity}$$
$$d(A, B) = d(B, A) \qquad \text{symmetry}$$
$$d(A, B) + d(B, C) \geq d(A, C) \qquad \text{triangle inequality.}$$

A measure of distance suitable for tree building should also have the *additivity* property, where additivity means that if we have an evolutionary event between three species (in two steps), A $\longrightarrow$ B $\longrightarrow$ C, i.e. A evolved into B and later B evolved into C, then the expected distances should add correctly, that is

$$d(A, B) + d(B, C) = d(A, C).$$

If the distances do not have the additivity property, no matter how much data we collect, we may reconstruct trees incorrectly.

Actual measures of distance are often subject to error. That is why we say that the equation should hold for the expected value.

There are three common measures of distance.

- **Percentage of changed positions**, equal to 100% minus the percentage of identity, a measure which is not additive.
- **Synteny**, a measure of the relative order of genes in the genomes which is very close to additive for close species.
- **PAM distance**, a measure of amino acid mutation which is additive.

How do we measure the distance and variance between two sequences? Let us get acquainted with percentage of changed positions and PAM distance using examples.

Example The sequence alignment of the protein triosephosphate isomerase in rice and human:

```
lengths=245,243 simil=1349.7, PAM_dist=52, identity=60.4

RKFFVGGNWKCNGTTDQVDKIVKILNEGQIASTDVVEVVVSPPYVFLPVVKSQLRPEIQVAAQNCWVK
|||||||||.||........:!..||..:!.:.  .|||.:||..!:...!.:|.|.|.|||||||:..
RKFFVGGNWKMNGRKQSLGELIGTLNAAKVPAD___TEVVCAPPTAYIDFARQKLDPKIAVAAQNCYKV

KGGAFTGEVSAEMLVNLSIPWVILGHSERRSLLGESNEFVGDKVAYALSQGLKVIACVGETLEQRESG
..||||||!|..|:..:....||!||||||||...|||:|.!|.|||:||:||.|||||!||.|!:|||
TNGAFTGEISPGMIKDCGATWVVLGHSERRHVFGESDELIGQKVAHALAEGLGVIACIGEKLDEREAG

STMDVVAAQTKAISERIKDWTNVVVAYEPVWAIGTGKVATPDQAQEVHDGLRKWLAANVSAEVAESTR
|..||  .|||.|:!.!|||:.||.||||||||||||.|||.|||||||!.||.||.:|||..||:|||
ITEKVVFEQTKVIADNVKDWSKVVLAYEPVWAIGTGKTATPQQAQEVHEKLRGWLKSNVSDAVAQSTR
```

```
IIYGGSVTGANCKELAAKPDVDGFLVGGASLKPEFIDIINS
|||||||||.|||||::||||||||||||||||!||||:
IIYGGSVTGATCKELASQPDVDGFLVGGASLKPEFVDIINA
```

The proteins are sequences of amino acids where each amino acid is identified by its one letter abbreviation, see Table 3.1. Since these sequences are longer than one line, the alignment is broken in parts. The sequences on the top are the rice sequences, the ones in the bottom are the human sequences. The rows in the middle are a visual indication of the quality of the alignment, where

| | means an exact match,
! means a very good match,
: means the match is ok,
. means the match is not good,
(blank) means a poor match or a gap.

The quality of the matches is derived from biochemical properties of the amino acids. Some are very similar, others are very different. Underscore characters indicate that one of the sequences is shorter, i.e. that a deletion in the sequence or an insertion in the other sequence has occurred.

Percentage of change

We could simply count the percentage of *identity* in the alignment. Our distance would be the percentage of mismatches:

number of exact matches	148
number of amino acids	245 and 243
length of the alignment	245
percentage of identity	$148/245 \approx 60.41\%$
percentage of change	$97/245 \approx 39.59\%$.

Then $1 - p$ (p stands for percent identity) is a measure of distance. This measure is very simple but has some problems. The main problem is that it does not satisfy the additivity property, since a point mutation during the change from amino acid sequence $A \to B$ which is reverted in the transition from amino acid sequence $B \to C$ will make $d(A, B) + d(B, C) > d(A, C)$.

The second problem is that percentage of change is a very crude measure of distance. In particular, when comparing amino acid sequences there is more information hidden in the mutations to different amino acids: some amino acids are chemically very similar, some are very different, so considering all the mutations as equal reduces the information.

Synteny

ABCDEFGHIJ

AEDCBFGHIJ

AEDCBHGFIJ

AEDCFGHBIJ

Three inversions in a gene sequence.

Synteny refers to the preservation of gene order in species descending from a common ancestor. Two species that are close will have the same or nearly the same order of genes. With time DNA rearrangements occur which will alter the order of genes in a genome. We model such a rearrangement with one or more inversions of a large portion of DNA, normally including many genes. The minimum number of inversions needed to reorder a genome of one species into the genome of another species can be used as a measure of distance, for example

> species 1 has its genes in the order ABCDEFGHIJ
> species 2 has its genes in the order AEDCFGHBIJ

Here each letter represents a gene – a long portion of DNA encoding one protein. These two sequences require three inversions to convert one into the other; as shown in Figure 8.7.

Hence we will say that the synteny distance between these two genomes is three. Genomes may have thousands of genes, so when there are few rearrangements the chance of one rearrangement exactly undoing a previous one is very small. Intuitively we can argue that each inversion, which is identified by the two end points, is rather unique: there are $\binom{n}{2}$ possible inversions for a genome with n genes. This is the reason why the synteny distance can be considered additive for practical purposes.

PAM distance

In the example alignment between rice and human the mismatches also tell us something. We can see that most of the differences are biased towards a good match. They cannot be considered random. So we need a theory, Markov model, that allows us to measure the distance between sequences using as much information as possible. Since this is a very important topic in bioinformatics and many other evolutionary processes, all the following Section 8.3.2 is devoted to it.

8.3.2 PAM distance in detail

To understand the concept of PAM distance we start with a Markovian model of evolution. This is a simplification of what happens in nature, yet it is quite powerful and the most widely accepted model of evolution.

A Markovian model implies that amino acids (AA) mutate randomly and independently of each other. Each AA mutates at a rate which depends only on itself and the resulting amino acid (and not on its history or its neighbors). Insertions or deletions also happen at random, independently of other events. Since there are 20 AAs, the transition probabilities are described by a 20×20 mutation matrix, denoted by $\mathbf{M}$ with entries M_{ij}, where[11]

$$M_{ji} = \text{probability \{amino acid } i \rightarrow \text{ amino acid } j\}.$$

Since an amino acid either mutates or stays the same, the sum of each column of $\mathbf{M}$ must be equal to one: $\sum_{j=1}^{20} M_{ji} = 1$. While simple, this is one of the best methods for modelling evolution.

A **1-PAM mutation matrix** describes an amount of evolution which will change, on the average, 1% of the amino acids. This is a very small amount of mutation, which means that the matrix is diagonally dominant; the diagonal entries are the largest ones (with probabilities of about 99%), see Table 8.2.

In mathematical terms this is expressed as a mutation matrix $\mathbf{M}$ such that

$$\sum_{i=1}^{20} f_i(1 - M_{ii}) = 0.01$$

where f_i is the frequency of the ith amino acid. The frequency of an AA in nature can be estimated by examining a large number of proteins, it is the total number of times this AA is found divided by the total number of examined AAs.

The vector $\mathbf{f} = (f_1, f_2, \ldots, f_{20})$ describes the frequencies of the amino acids in nature, consequently $\sum_{i=1}^{20} f_i = 1$ and

$$\sum_{i=1}^{20} f_i(1 - M_{ii}) = 1 - \sum_{i=1}^{20} f_i M_{ii}.$$

The M_{ii} values, the diagonal elements of $\mathbf{M}$, are the probabilities that a given AA does not change. The name "PAM" is an acronym for "point accepted mutations," which is only of historical value. For us a PAM unit of distance is an amount of mutation defined by the matrix $\mathbf{M}$ as above.

---

[11] We are following the original notation of Dayhoff. In the Markovian literature there is a tendency to work with the transpose of the matrix $\mathbf{M}$. Other than the suitable transpositions there is no difference.

Table 8.2 The 1-PAM mutation matrix **M** (all values multiplied by 10 000)

	A	R	N	D	C	Q	E	G	H	I	L	K	M	F	P	S	T	W	Y	V
A	9890	5	5	6	12	9	11	12	5	2	5	6	9	2	10	29	14	1	2	17
R	4	9907	5	2	2	16	4	3	8	1	2	30	2	0	3	5	5	4	3	2
N	3	4	9888	18	2	8	5	6	13	1	1	10	1	1	2	13	8	1	3	1
D	4	2	21	9905	0	7	28	5	6	0	0	5	0	0	3	7	5	0	1	0
C	3	1	1	0	9946	0	0	1	1	1	1	0	1	1	0	3	1	1	1	2
Q	4	11	7	5	1	9856	18	2	14	1	3	14	6	1	4	5	5	1	1	2
E	8	5	6	30	0	28	9890	2	7	1	1	15	3	0	4	7	5	1	1	3
G	11	4	9	7	2	4	3	9952	3	0	1	3	1	0	2	10	2	2	1	1
H	1	4	7	3	1	9	3	1	9895	1	1	3	2	2	1	2	2	1	9	1
I	2	1	2	0	2	2	1	0	2	9878	22	2	26	7	1	1	5	2	2	42
L	5	4	2	0	3	8	2	1	3	35	9919	3	48	22	4	3	4	5	5	19
K	5	33	13	5	0	22	15	2	8	2	2	9883	5	1	4	6	9	1	2	2
M	3	1	1	0	2	4	1	0	2	10	12	2	9859	5	0	2	3	1	1	4
F	1	0	1	0	3	1	0	0	4	5	10	0	9	9923	0	1	1	10	28	3
P	6	2	2	2	0	5	3	1	2	1	2	3	0	1	9943	6	5	0	1	2
S	23	5	17	8	9	9	7	8	6	1	2	7	4	1	8	9862	32	2	4	2
T	11	5	11	6	4	8	5	2	7	6	2	9	7	2	7	33	9879	1	2	12
W	0	1	0	0	1	1	0	0	1	0	1	0	1	3	0	0	0	9956	4	0
Y	1	2	2	1	3	1	1	0	13	1	2	1	2	22	1	2	1	10	9924	2
V	15	2	1	0	8	3	4	1	2	51	14	3	12	5	2	3	14	1	4	9884

If we have an amino acid probability vector p, the product $\mathbf{M}p$ gives the probability vector after a random evolution, equivalent to a 1-PAM unit. Or, if we start with a given AA i (i.e. a probability vector which contains a 1 in position i and zeros in all other positions), $\mathbf{M}p = M_{*i}$ (the ith column of $\mathbf{M}$) is the corresponding probability vector after one unit of evolution.

After two consecutive PAM units we will have $\mathbf{M}(\mathbf{M}p) = \mathbf{M}^2 p$. Similarly, after k consecutive units of evolution (sometimes called k-PAM evolution) a probability vector p will be changed into the probability vector $\mathbf{M}^k p$. This reasoning establishes the relation between mutation matrices and PAM distances, so for a particular evolutionary process, all mutation matrices (corresponding to different distances) can be expressed as powers of a unit mutation matrix raised to the distance, for example $\mathbf{M}^k$.

After infinite evolution, any probability distribution converges to the distribution in nature, i.e. the natural frequency vector f. Consequently $\mathbf{M}^\infty p = f$. Since f is the limiting value, it is not affected by further mutation, or $\mathbf{M}f = f$. In other words, f is the eigenvector with eigenvalue 1 of $\mathbf{M}$.

A k-**PAM mutation matrix** is given by $\mathbf{M}^k$ (see Table 8.3). Two sequences which are 250 PAM units distant from each other are expected to have approximately 17%

Table 8.3 The 250-PAM mutation matrix $\mathbf{M}^{250}$ (all values multiplied by 10 000), note that the matrix is weakly diagonally dominant

	A	R	N	D	C	Q	E	G	H	I	L	K	M	F	P	S	T	W	Y	V
A	1350	677	733	733	884	748	781	884	647	649	597	714	671	461	834	1006	899	342	468	803
R	460	1583	568	493	323	751	589	424	616	304	320	995	361	253	429	512	507	366	351	337
N	425	485	1092	747	299	534	563	496	601	239	228	552	275	225	366	558	507	197	326	271
D	501	496	881	1593	260	666	1007	549	591	221	213	604	268	188	453	597	536	164	284	273
C	212	114	124	91	2660	107	94	119	138	145	134	100	153	157	93	193	166	149	169	187
Q	359	530	442	468	215	705	558	300	496	243	261	538	303	208	359	390	379	201	253	264
E	580	645	722	1095	293	863	1334	483	635	315	305	768	370	237	525	614	572	215	312	376
G	827	583	801	752	466	585	608	3387	530	264	264	567	332	224	512	799	569	291	292	346
H	194	271	310	259	173	309	256	170	946	143	153	267	175	231	182	225	221	194	387	149
I	480	330	304	239	446	375	314	208	353	1460	1094	354	1027	726	320	379	510	381	489	1185
L	717	566	473	374	673	653	492	339	612	1779	2390	576	1798	1501	561	575	707	793	942	1443
K	535	1097	713	662	311	840	774	454	667	358	359	1240	425	276	510	600	605	264	360	394
M	195	154	138	114	185	183	144	103	170	403	435	165	608	326	130	166	198	181	217	330
F	239	193	202	143	340	224	165	124	399	510	649	191	583	2041	171	213	246	937	1312	419
P	484	367	366	385	225	434	410	318	353	251	271	396	259	191	2614	495	467	144	222	299
S	774	580	741	671	619	625	636	657	578	394	368	616	439	315	656	997	844	285	393	475
T	707	587	687	616	544	621	605	478	580	542	462	635	535	372	632	862	1101	274	397	622
W	57	90	57	40	103	70	48	52	108	86	110	59	104	301	41	62	58	3398	336	72
Y	195	215	235	173	293	220	175	130	539	276	327	200	311	1054	160	213	211	843	1955	253
V	708	437	413	352	689	486	446	326	440	1415	1060	464	1004	712	454	545	698	382	535	1501

identity (1 out of 6) for the most common AA mutation matrices, i.e.

$$\sum_{i=1}^{20} f_i \left(\mathbf{M}^{250}\right)_{ii} \approx 0.17.$$

We say that two sequences are homologous when they have evolved from a common ancestor. PAM distances of 250 are the practical limit of detection for homology. Well below 250, the similarity is obvious, much above 250 it is likely to be impossible to establish.

Example 1

```
CNGTTDQVDKIVKILNEGQIASTDVVEVVVSPPYVFLPVVKSQLRPEIQV
|||||||||||||||:||||||||||||||||||||||||||||||||||
CNGTTDQVDKIVKIRNEGQIASTDVVEVVVSPPYVFLPVVKSQLRPEIQV
lengths = 50, 1 Mismatch, PAM distance ≈ 2
```

Example 2

```
CNGTKESITKLVSDLNSATLEAD--VDVVVAPPFVYIDQVKSSLTGRVEISAQNCWIGKG
||||.!.!.|:|.  ||...:.:.   |!|||:|||!|!:..|||.|...!:!:|||||!.||
CNGTTDQVDKIVKILNEGQIASTDVVEVVVSPPYVFLPVVKSQLRPEIQVAAQNCWVKKG
lengths = 58 and 60, PAM distance ≈ 83
```

An explanation of the encoding of the middle row can be found on page 151.

Definition of PAM distance The distance between two aligned sequences is defined as the exponent k of the matrix $\mathbf{M}$ such that the joint probability of all mutations is maximized for $\mathbf{M}^k$. This is by definition a maximum likelihood estimate. Defined in this way PAM distances are additive, but the proof of this statement goes beyond the scope of this section.

PAM distances do not necessarily correspond to time! PAM distances measure the amount of change, of mutation, and the mutation rate does not depend on time alone, but on many other important factors, for example the following.

- The accuracy of DNA reproduction, which is highly dependent on the life form. Eukaryotes using internal correction mechanisms have a far higher accuracy of DNA reproduction than e.g. bacteria. The more accurate DNA reproduction is, the slower is the rate of change.
- The time needed for a replication cycle. This generation time varies from minutes for bacteria to hundreds of years in redwood trees.
- The number of genome duplications between generations. One for unicellular organisms, about 20 for mammalian females, 60 for mammalian males.
- The amount of functional conservation in the case of coding DNA. Proteins that have been extensively optimized show functional conservation: they are more intolerant to changes, i.e. natural selection will disfavor changes.
- Changes in the environment that will affect the protein. Random mutations are more likely to be beneficial under such conditions, hence speeding up the mutation rate.
- The need to survive or resist attack. As above, random mutations are more likely to be beneficial, hence speeding up the mutation rate.

As a consequence, PAM distances measure the amount of mutation well, but can be inaccurate when measuring time.

The mutation matrices have a symmetry property for any distance k. If we think of the state transition diagrams for amino acid i and amino acid j in steady state, this means that the total flow from i to j (the probability of being in state i and mutating to state j) is the same as the total flow from j to i (Figure 8.8).

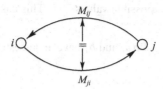

Figure 8.8   The symmetry of the state transitions in steady state.

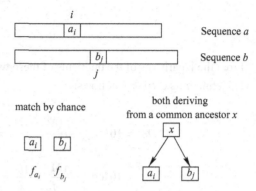

Figure 8.9   Match by chance versus match by common ancestry.

This implies a simple relation for $\mathbf{M}^k$:

$$f_i \left(\mathbf{M}^k \right)_{ji} = f_j \left(\mathbf{M}^k \right)_{ij}.$$

The mutation matrices which we use have this symmetry property for any distance k.

8.3.3 Calculating distances from sequence alignment

Aligning sequences has as goal finding the alignment which maximizes the probability that the two sequences evolved from a common ancestral sequence, as opposed to being random sequences (Figure 8.9). More precisely, for each pair of amino acids, which we align, we are considering two possible events.

Case 1 The two positions are independent of each other, and hence an arbitrary position with a_i is aligned to another arbitrary position with b_j. This has a probability equal to the product of the individual frequencies

$$\Pr(\text{independent alignment of } a_i \text{ and } b_j) = f_{a_i} f_{b_j}.$$

Case 2 The two positions have evolved from a common ancestor x after t_a and t_b units of evolution. We assume that the common ancestor has an unknown amino acid x at the position we want to align. Since x is unknown, we have to sum for all possible values of x. This has a probability

$$\Pr(a_i \text{ and } b_j \text{ have an ancestor } x) = \sum_x f_x \Pr\{x \rightarrow a_i\} \Pr\{x \rightarrow b_j\}$$

$$= \sum_x f_x \left(\mathbf{M}^{t_a}\right)_{a_i x} \left(\mathbf{M}^{t_b}\right)_{b_j x}$$

$$= \sum_x f_{b_j} \left(\mathbf{M}^{t_a}\right)_{a_i x} \left(\mathbf{M}^{t_b}\right)_{x b_j}$$

$$= f_{b_j} \left(\mathbf{M}^{t_a+t_b}\right)_{a_i b_j} = f_{a_i} \left(\mathbf{M}^{t_a+t_b}\right)_{b_j a_i}.$$

From the logarithms of the quotients of these two probabilities we can build a matrix **D**, defining its entries as follows:

$$D_{ij} = 10 \log_{10} \frac{f_j \left(\mathbf{M}^{t_a+t_b}\right)_{ij}}{f_i f_j} = 10 \log_{10} \frac{\left(\mathbf{M}^{t_a+t_b}\right)_{ij}}{f_i}$$

$$= 10 \log_{10} \frac{f_i (\mathbf{M}^{t_a+t_b})_{ji}}{f_i f_j} = 10 \log_{10} \frac{\left(\mathbf{M}^{t_a+t_b}\right)_{ji}}{f_j}.$$

Notice that **D** is symmetric – the last term in the first line is equal to the last term in the second line. It is easy to see that the matrix **D** depends on the total distance between a and b and not the individual distances t_a and t_b.

8.3.4 Dayhoff matrices

In the standard dynamic programming method of sequence alignment, Dayhoff matrices[12] **D** are used for scoring. The matrix entries of the Dayhoff matrix **D** are derived from the 250-PAM mutation matrix $\mathbf{M}^{250}$ (see Table 8.4) by

$$D_{ij} = 10 \log_{10} \frac{\left(\mathbf{M}^{250}\right)_{ij}}{f_i} \quad \text{(see above)}.$$

From this point onwards, we shall work with Dayhoff matrices.

Notice that there is a large "penalty" (negative score) for exchanging some amino acids. This normally means that the pairs of amino acids are chemically and/or

---

[12] Named in honor of Margaret O. Dayhoff, who introduced these matrices in M. O. Dayhoff, R. M. Schwartz, and B. C. Orcutt, A model of evolutionary change in proteins, in: *Atlas of Protein Sequence and Structure*, National Biomedical Research Foundation, 1978. Any logarithm (and any factor of it) will give similar results. $10 \log_{10}$ was used originally and is the most common in use.

Table 8.4 A Dayhoff matrix derived from $\mathbf{M}^{250}$, $\mathbf{D}$ is symmetric

C	11.5																			
S	0.1	2.2																		
T	−0.5	1.5	2.5																	
P	−3.1	0.4	0.1	7.6																
A	0.5	1.1	0.6	0.3	2.4															
G	−2.0	0.4	−1.1	−1.6	0.5	6.6														
N	−1.8	0.9	0.5	−0.9	−0.3	0.4	3.8													
D	−3.2	0.5	−0.0	−0.7	−0.3	0.1	2.2	4.7												
E	−3.0	0.2	−0.1	−0.5	−0.0	−0.8	0.9	2.7	3.6											
Q	−2.4	0.2	0.0	−9.2	−0.2	−1.0	0.7	0.9	1.7	2.7										
H	−1.3	−0.2	−0.3	−1.1	−0.8	−1.4	1.2	0.4	0.4	1.2	6.0									
R	−2.2	−0.2	−0.2	−0.9	−0.6	−1.0	0.3	−0.3	0.4	1.5	0.6	4.7								
K	−2.8	0.1	0.1	−0.6	−0.4	−1.1	0.8	0.5	1.2	1.5	0.6	2.7	3.2							
M	−0.9	−1.4	−0.6	−2.4	−0.7	−3.5	−2.2	−3.0	−2.0	−1.0	−1.3	−1.7	−1.4	4.3						
I	−1.1	−1.8	−0.6	−2.6	−0.8	−4.5	−2.8	−3.8	−2.7	−1.9	−2.2	−2.4	−2.1	2.5	4.0					
L	−1.5	−2.1	−1.3	−2.3	−1.2	−4.4	−3.0	−4.0	−2.8	−1.6	−1.9	−2.2	−2.1	2.8	2.8	4.0				
V	−0.0	−1.0	0.0	−1.8	0.1	−3.3	−2.2	−2.9	−1.9	−1.5	−2.0	−2.0	−1.7	1.6	3.1	1.8	3.4			
F	−0.8	−2.8	−2.2	−3.8	−2.3	−5.2	−3.1	−4.5	−3.9	−2.6	−0.1	−3.2	−3.3	1.6	1.0	2.0	0.1	7.0		
Y	−0.5	−1.9	−1.9	−3.1	−2.2	−4.0	−1.4	−2.8	−2.7	−1.7	2.2	−1.8	−2.1	−0.2	−0.7	−0.0	−1.1	5.1	7.8	
W	−1.0	−3.3	−3.5	−5.0	−3.6	−4.0	−3.6	−5.2	−4.3	−2.7	−0.8	−1.6	−3.5	−1.0	−1.8	−0.7	−2.6	3.6	4.1	14.2
	C	S	T	P	A	G	N	D	E	Q	H	R	K	M	I	L	V	F	Y	W

biologically very different, e.g.

$$L \leftrightarrows G = -4.4$$
$$I \leftrightarrows G = -4.5$$
$$W \leftrightarrows P = -5.0$$
$$F \leftrightarrows G = -5.2.$$

Some amino acids tend not to mutate (large positive score with themselves), which means that these amino acids have specialized functions, so they are difficult to exchange, e.g.

$$W \leftrightarrows W = 14.2$$
$$C \leftrightarrows C = 11.5$$
$$P \leftrightarrows P = 7.6,$$

while other amino acids have a low value for remaining the same, indicating that they are easily exchanged, e.g.

$$S \leftrightarrows S = 2.2$$
$$A \leftrightarrows A = 2.4.$$

Other amino acids seem to mutate easily into each other, which means that they are relatively similar, e.g.

$$Y \leftrightarrows F = 5.1$$
$$W \leftrightarrows Y = 4.1$$
$$V \leftrightarrows I = 3.1$$
$$L \leftrightarrows M = 2.8.$$

Most amino acids mutate with little bonus or penalty for changing:

$$P \leftrightarrows T = 0.1$$
$$V \leftrightarrows T = 0.0$$
$$V \leftrightarrows C = 0.0$$
$$D \leftrightarrows T = 0.0.$$

It should be remembered that PAM-250 is a very large amount of evolution and the above comments make sense in view of this extreme situation.

8.3.5 Sequence alignment using Dayhoff matrices and maximum likelihood

The alignment of sequences is done with dynamic programming which uses a scoring matrix to score each pair of amino acids and maximizes the sum of scores. The sums to be maximized are the sum of scores of the individual amino acids matched plus the penalties for any gaps needed. If we use Dayhoff matrices for scoring individual AAs, we find the following: let a_i and b_i be the amino acids of two sequences which are at position i of the alignment. Then, ignoring insertions/deletions, the alignment score is:

$$\sum_i D_{a_i b_i} = \sum_i 10 \log_{10} \left(\frac{\Pr(a_i \text{ and } b_i \text{ have a common ancestor})}{\Pr(a_i \text{ and } b_i \text{ are random})} \right)$$
$$= 10 \log_{10} \frac{\prod_i \Pr(a_i \text{ and } b_i \text{ have a common ancestor})}{\prod_i \Pr(a_i \text{ and } b_i \text{ are random})}.$$

Under the assumption that positions evolve independently of each other, the product in the numerator represents the joint probability of the two entire sequences evolving from a common ancestor, and the product in the denominator represents the probability of the two sequences being randomly aligned. The bottom probability does not depend on the alignment, it is the product of all individual frequencies and hence it is constant for any two given sequences. The top part depends on the alignment.

Hence when we maximize the score using Dayhoff matrices, we find the alignment which maximizes the likelihood of coming from a common ancestor, i.e. we have a maximum likelihood alignment.

This makes aligning sequences using Dayhoff matrices a soundly based algorithm, better founded than any other alignment algorithm. Recall that maximum likelihood estimates are

- unbiased (with infinite information they give the correct results),
- asymptotically normal, and
- most powerful (in the sense of having the smallest variance of the estimator).

How to score insertions/deletions

How do we account for insertions/deletions when calculating the distance of alignments? First we note that there is a symmetry between insertions and deletions. Since we do not know the ancestor sequences, it is generally impossible to conclude whether the ancestor suffered a deletion in one of the branches or an insertion into the other. Hence we usually call these events "indels" and we talk about the probability of an indel, e.g.

```
...AKPQLLT...
...AN__ILS...
```

Here we can argue that PQ was inserted into the top sequence or that PQ was deleted from the bottom sequence. As we did with the pairing of amino acids, we want to compute the score from the ratio of the probabilities of two events:

(i) no relation, i.e. two random sequences (shown next left) and
(ii) both sequences evolved from the same ancestor (shown next right)

We will analyze the given partial alignment as a composite event:

No relation		Common ancestor		
A	A	A	$\leftarrow X_1 \rightarrow$	A
K	N	K	$\leftarrow X_2 \rightarrow$	N
P	I	P	inserted	
Q	L	Q	inserted	
L	S	L	$\leftarrow X_3 \rightarrow$	I
L		L	$\leftarrow X_4 \rightarrow$	L
T		T	$\leftarrow X_5 \rightarrow$	S

If there is no relation (and hence no common ancestor X) the event has as probability the product of all the frequencies:

$$\Pr(\overline{X}) = (f_A \cdot f_K \cdot f_P \cdot f_Q \cdot f_L^2 \cdot f_T) \cdot (f_A \cdot f_N \cdot f_I \cdot f_L \cdot f_S).$$

Note: the random development of `ANILS` has exactly the same probability as the development of `NLISA`. The order of the AAs does not play any role.

The probability of the sequences having X as a common ancestor is:

$$\Pr(X) = f_A M_{AA} \cdot f_K M_{NK} \cdot \Pr(\text{indel}(2)) \cdot f_P \cdot f_Q \cdot f_L M_{IL} \cdot f_L M_{LL} \cdot f_T M_{ST},$$

where $\Pr(\text{indel}(2))$ is the probability of an indel of length 2.

This formula comes from the use of the pair of amino acids formula (see above) applied to the five pairs and for the event of an indel of length 2 times the natural individual frequencies of the amino acids inserted. So the ratio to maximize is (re-arranged conveniently)

$$\frac{\Pr(X)}{\Pr(\overline{X})} = \frac{f_A M_{AA}}{f_A f_A} \cdot \frac{f_K M_{NK}}{f_K f_N} \cdot \frac{\Pr(\text{indel}(2)) f_P f_Q}{f_P f_Q} \cdot \frac{f_L M_{IL}}{f_L f_I} \cdot \frac{f_L M_{LL}}{f_L f_L} \cdot \frac{f_T M_{TS}}{f_T f_S}$$

and taking $10 \log_{10}$ we get:

$$D_{AA} + D_{NK} + 10 \log_{10} \Pr(\text{indel}(2)) + D_{IL} + D_{LL} + D_{ST}.$$

We can generalize from this example that the cost of an indel of length k should be

$$\text{Cost}(\text{indel}(k)) = 10 \log_{10} \Pr(\text{indel}(k)).$$

This gives a precise basis to the assignment of indel costs. For example, we can collect large numbers of alignments and find the actual frequencies of indels which lets us estimate their probabilities and hence calculate their costs for dynamic programming.

For example, if we collect alignments of sequences (which we can trust as correct) and we have a total of 10 000 amino acids aligned and have found 31 indels of length 1, then the cost of an indel of length 1 should be $10 \log_{10} 31/10\,000 \approx -25.09$.

8.3.6 Estimating distances between sequences by maximum likelihood

The Dayhoff matrices and hence the scores (logarithms of probabilities) depend on the PAM distance used for the mutation matrix M. Consequently we can choose the PAM distance for M which will maximize the score. As we know, maximum likelihood estimates are unbiased and best asymptotic estimates. Because it is unbiased, this estimate of the distance is additive for sufficiently long sequences.

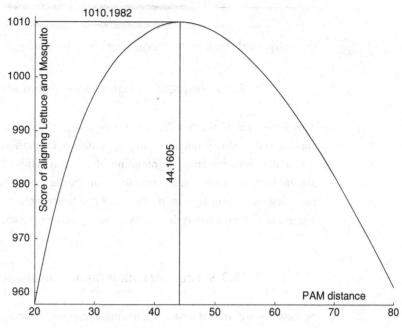

Figure 8.10   PAM distance versus score.

Example

```
Lettuce (lactuca sativa), Mosquito (culex pipens)
lengths=165,165 simil=1010.2, PAM_dist=44.0217, identity=63.6%
```

```
VAAQNCWVKKGGAFTGEVSAEMLANLGVPWVILGHSERRALLNETNEFVGDKVAYALSQGLKVIACVGE
||||||:....|||||||!|..||.:|.! |||||||||||:..|:.|.!.||!.:||::|||||||!||
VAAQNCYKVAKGAFTGEISPAMLKDLNIGWVILGHSERRAIFGESGELIADKIVHALAEGLKVIACIGE

TLEQREAGTTMEVVAAQTKAIADKISSWDNVVLAYEPVWAIGTGKVASPAQAQEVHAGLRKWFCDNVSA
||::|||..|..|. .|||||||||!..|.|||:||||||||||||.|:|.|||||||.||||.!||||
TLQEREAAKTEAVCFRQTKAIADKVKDWSNVVIAYEPVWAIGTGKTATPEQAQEVHAALRKWFTENVSA

EVSASTRIIYGGSVSGSNCKELGGQTD
!||::.|| |||||:.:||!||...:.|
DVSSAIRIQYGGSVTAANCRELAAKPD
```

The score for the alignment of lettuce and mosquito at different PAM distances is displayed in Figure 8.10. It can be seen that the maximum score is achieved at around 44 PAM. Using the second derivative of the logarithm of the likelihood we can estimate the variance of this PAM distance.

Figure 8.11 Local alignment (thick parts of sequences are aligned, thin parts are ignored).

8.3.7 Sequence alignment – global alignment

This is also called the Needleman–Wunsch algorithm. It is dynamic programming applied to the entire sequences using the scores of Dayhoff matrices. See Section 7.6 for full details of dynamic programming. Notice that in this case the Dayhoff matrices are similarity matrices, i.e. we want to maximize the score. (In Section 7.7.3 on string matching we were using penalty functions and hence were minimizing.) Aligning the sequences in their entirety is usually called "global alignment."

8.3.8 Sequence alignment – local alignment

Sometimes we are interested in finding the subsequences of the input sequences which align at the highest score. In other words we are ready to ignore any deletions which happen at the beginning or end of any of the two sequences. This is called *local alignment* or the Smith–Waterman algorithm (Figure 8.11). There are good biological reasons for using this type of alignment, in particular when we want to search a portion of a sequence which could be immersed in another.

Local alignments can be computed with scores arising from Dayhoff matrices (similarity scores), when these are maximized, but *not* with editing distances or penalty functions (see minimizing, Section 7.7.3).

The mechanism for computing a local alignment is quite simple.

(i) Compute the cost matrix as in Needleman–Wunsch.
(ii) Replace any negative entry (while computing it) with zero.
(iii) To start backtracking, choose the highest entry in the whole matrix. (This position will mark the end of the alignment.)
(iv) Backtrack up to the upper left corner or up to a zero, whichever comes first. (This position will mark the beginning of the alignment.)

8.3.9 Sequence alignment – cost-free end alignment

A third variant of alignments is called cost-free end alignment (CFE). CFE is a method which can be viewed as in between global and local alignment. In CFE we do not penalize for a single insertion/deletion (indel) at the beginning or at the end of the sequences. If we penalized for end-indels in both sequences, this would be a global alignment. If we do not penalize for any end-indel then this would be a local alignment. Since CFE penalizes for one end-indel on each side, it can be viewed as an intermediate algorithm between global and local. This is computed easily with a small variation of the global alignment.

(i) Make the top row and leftmost column entries all zero.
(ii) Compute the cost matrix as in Needleman–Wunsch.
(iii) Choose the largest value in the bottom row or rightmost column to start the backtracking. (This position will mark the end of the alignment.)
(iv) Backtrack up to the top row or leftmost column. (This will mark the beginning of the alignment.)

8.3.10 Distance and variance matrices

When we have a set of n homologous sequences (all having a common ancestor) we can align each pair and compute its distance (and also the variance of this distance). We normally place the distances in an $n \times n$ matrix which we call a *distance matrix*.

A distance matrix

	Lettuce	Rice	Mosquito	Monkey	Human
Lettuce	0.000	18.409	44.161	44.287	44.398
Rice	18.409	0.000	46.621	51.570	51.683
Mosquito	44.161	46.621	0.000	40.733	40.833
Monkey	44.287	51.570	40.733	0.000	0.804
Human	44.398	51.683	40.833	0.804	0.000

A distance matrix between n sequences is a symmetric $n \times n$ matrix, where the ij entry contains the PAM distance between sequences i and j, usually computed from the alignment of sequences i and j. Naturally the diagonal entries are zero.

It is also possible to compute the variances of the distances and hence get a measure of their precision. This information is very useful for building phylogenetic trees.

A variance matrix

	Lettuce	Rice	Mosquito	Monkey	Human
Lettuce	0.000	10.621	35.460	32.296	32.450
Rice	10.621	0.000	33.808	30.669	30.799
Mosquito	35.460	33.808	0.000	28.614	28.748
Monkey	32.296	30.669	28.614	0.000	0.323
Human	32.450	30.799	28.748	0.323	0.000

8.4 Creating an initial topology

We will use a simple approximation algorithm to create an initial tree. The method of choice is a greedy heuristic, which joins at each step the two closest leaves or subtrees which are not already joined.[13] This is a special case of the nearest neighbor heuristics which are used for many problems. If we ignore the weights this is called UPGMA (unweighted pair group method with arithmetic mean). We shall introduce this widely used method next and consider the improved version using the weights (WPGMA) in Section 8.4.2.

8.4.1 The UPGMA algorithm

As an example we use the sequence data of *Lactuca sativa* (garden lettuce) L, *Oryza sativa* (rice) R, *Culex pipens* (house mosquito) P, *Macaca mulata* (Rhesus monkey) M, and *Homo sapiens* (human) H.

The initial tree will be built from the PAM distances given above in Section 8.3.10.

The minimum distance is 0.804 for human–monkey. So we will join H and M, human and monkey, into a new subtree A, the subtree with the two leaves M and H (Figure 8.12). For the rest of the construction this subtree will not change. Hence H and M will not be used again and the root of A becomes a new node to join like a new leaf.

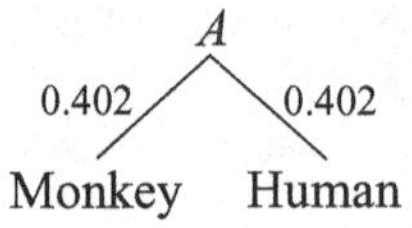

Figure 8.12 The initial tree for monkey and human using UPGMA.

After we have joined the two species in a subtree we have to compute the distances from every other node to the root of the new subtree. We do this with a simple average of distances:

$$d(L, A) = \frac{d(L, M) + d(L, H)}{2} - \frac{d(H, M)}{2} = 43.941$$

$$d(R, A) = \frac{d(R, M) + d(R, H)}{2} - \frac{d(H, M)}{2} = 51.225$$

$$d(P, A) = \frac{d(P, M) + d(P, H)}{2} - \frac{d(H, M)}{2} = 40.381.$$

---

[13] A greedy algorithm is defined as an algorithm which optimizes the immediate next step, as opposed to optimizing the entire procedure. In general, greedy algorithms are fast but not optimal.

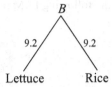

Figure 8.13   The initial subtree for lettuce and rice using UPGMA.

After the first step we have the following PAM distances:

	Lettuce L	Rice R	Mosquito P	A
Lettuce L	0.0	**18.409**	44.161	43.941
Rice R	**18.409**	0.0	46.621	51.225
Mosquito P	44.161	46.621	0.0	40.381
A	43.941	51.225	40.381	0.0

For the second step, $d(L, R) = 18.409$ is the minimum. So we join L and R, let-tuce and rice, introducing a root in the middle, resulting in the subtree shown in Figure 8.13.

Again we compute the average of distances and get:

$$d(B, A) = \frac{d(L, A) + d(R, A)}{2} - \frac{d(L, R)}{2} = 38.379$$

$$d(B, P) = \frac{d(L, P) + d(R, M)}{2} - \frac{d(L, R)}{2} = 36.186.$$

As result we get the following PAM distances:

	B	Mosquito P	A
B	0.0	**36.186**	38.379
Mosquito P	**36.186**	0.0	40.381
A	38.379	40.381	0.0

Now $d(B, P) = 36.186$ is the minimum. So we join B and P (mosquito) to one subtree with root C by averaging the distances:

$$\frac{1}{2}d(B, P) = 18.093 = d(C, B) = d(C, P)$$

$$d(C, A) = \frac{d(B, A) + d(P, A)}{2} - \frac{(B, P)}{2} = 21.287,$$

and we get the following PAM matrix

	C	A
C	0.0	21.287
A	21.287	0.0

The last step is trivial, we have the distance between the two nodes A and C (Figure 8.14).

As presented, the UPGMA algorithm will resolve ultrametric trees exactly. An ultrametric tree is a tree where the distances are exactly clock-like, as if all the proteins evolved at exactly the same rate. For more on ultrametric trees see Section 8.6.2.

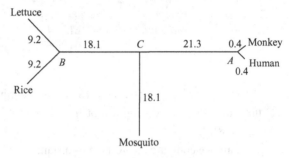

Figure 8.14 The complete unrooted tree for our five species using UPGMA.

8.4.2 The WPGMA algorithm

The weighted version of the previous algorithm, weighted pair group method with arithmetic mean (WPGMA), has two improvements. The first one is to use the inverses of the variances as weights and the second is to optimize the construction of the subtrees by choosing the branches of new subtrees according to the distances to the rest of the leaves instead of arbitrarily halving it.

Weighted averages

To compute a weighted average, we use the formula:

$$\text{Ave}(x_1, x_2, \ldots, x_n) = \frac{\sum_{i=1}^{n} w_i x_i}{\sum_{i=1}^{n} w_i},$$

and the variance of the average is

$$\sigma^2(\text{Ave}(x_1, x_2, \ldots, x_n)) = \frac{\sum_{i=1}^{n} w_i^2 \sigma_{x_i}^2}{(\sum_{i=1}^{n} w_i)^2}.$$

When we have the variances of our distances, the correct weight to use is $w_i = 1/\sigma^2_{x_i}$, so our average formula becomes

$$\mathrm{Ave}\,(x_1, x_2, \ldots, x_n) = \frac{\displaystyle\sum_{i=1}^{n} \frac{x_i}{\sigma^2_{x_i}}}{\displaystyle\sum_{i=1}^{n} \frac{1}{\sigma^2_{x_i}}},$$

and the variance of the average simplifies to

$$\sigma^2(\mathrm{Ave}) = \frac{1}{\displaystyle\sum_{i=1}^{n} \frac{1}{\sigma^2_{x_i}}}.$$

The WPGMA algorithm applied to our example

Figure 8.15

The initial tree for monkey and human using WPGMA.

We compute the weighted distance from monkey and human to the other three species:

$$d_{\mathrm{M}} = \frac{d(M, L)}{\sigma^2(M, L)} + \frac{d(M, R)}{\sigma^2(M, R)} + \frac{d(M, P)}{\sigma^2(M, P)} = 45.437$$

$$d_{\mathrm{H}} = \frac{d(H, L)}{\sigma^2(H, L)} + \frac{d(H, R)}{\sigma^2(H, R)} + \frac{d(H, P)}{\sigma^2(H, P)} = 45.546.$$

The new subtree will have two leaves which are separated by 0.804. So the two branches from the root A to the leaves M and H should add to 0.804 while their difference should be $d_{\mathrm{M}} - d_{\mathrm{H}} = 45.437 - 45.546 = -0.109$. This results in the subtree in Figure 8.15.

The new distance from A to lettuce L is

$$\frac{\frac{44.398 - 0.456}{32.450} + \frac{44.287 - 0.347}{32.296}}{\frac{1}{32.450} + \frac{1}{32.296}} = 43.941$$

and its variance is

$$\frac{1}{\frac{1}{32.450} + \frac{1}{32.296}} = 16.187.$$

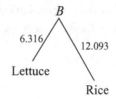

Figure 8.16   The initial tree for lettuce and rice.

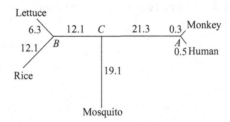

Figure 8.17   The complete unrooted tree for our five species using WPGMA.

Computing the new distances and variances we obtain the following.

PAM distance

	Lettuce L	Rice R	Mosquito P	A
Lettuce L	0.0	**18.409**	44.161	43.941
Rice R	**18.409**	0.0	46.621	51.225
Mosquito P	44.161	46.621	0.0	40.381
A	43.941	51.225	40.381	0.0

PAM variance

	Lettuce L	Rice R	Mosquito P	A
Lettuce L	0.0	10.621	35.460	16.187
Rice R	10.621	0.0	33.808	15.367
Mosquito P	35.460	33.808	0.0	14.340
A	16.187	15.367	14.340	0.0

Now the closest nodes are L and R at PAM distance 18.409. We join them using all available information, as shown in Figure 8.16.

And after two more steps we obtain the tree shown in Figure 8.17.

The WPGMA is an excellent approximate constructor. See also the interactive exercise "Language."

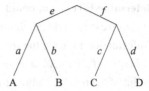

Figure 8.18   A simple tree with four leaves.

8.4.3 Tree fitting by least squares

When we have distance information we store it in a distance matrix $\mathbf{D}$. For a particular tree, we can measure the distances over the tree and call them T_{ij}. An example for a very simple tree with four leaves is shown in Figure 8.18.

The T_{ij} give an $n \times n$ matrix $\mathbf{T}$. Notice that $\mathbf{T}$ depends on the branch lengths but also on the topology of the tree. The tree construction problem is defined as finding a tree (topology and branch lengths) such that $\|\mathbf{D} - \mathbf{T}\|$ is minimized. If we use the common euclidean norm for measuring the value of $\|\mathbf{D} - \mathbf{T}\|$ we have a method which is called *least square trees*. More precisely we have to minimize

$$\|\text{error}\|^2 = \|\mathbf{D} - \mathbf{T}\|^2 = \sum_{i,j}(D_{ij} - T_{ij})^2$$

for the distances and

$$\|\text{error}\|^2 = \sum_{i,j} \frac{(D_{ij} - T_{ij})^2}{\sigma_{ij}^2},$$

for the variances of the distances.

For a fixed topology, finding the optimal branch lengths becomes a problem of linear least squares:

$$
\begin{aligned}
T_{A,B} &= \quad a + b \quad &\cong \text{dist[A, B]} \\
T_{A,C} &= a + e + f + c &\cong \text{dist[A, C]} \\
&\;\;\vdots \\
T_{C,D} &= \quad c + d \quad &\cong \text{dist[C, D]}.
\end{aligned}
$$

Notice that if the tree is the one with the correct topology, then, for example, the distance between the species C and D is measured by the edge lengths $c + d$. We have an approximation of this value given by dist[C, D]. So the variable $(c + d - \text{dist[C, D]})/\sigma^2(C, D)$ is a random variable with expected value zero (if the distance estimate is unbiased) and variance one. Consequently it makes perfect sense to minimize the quadratic norm of these errors. With the least squares computation of edge lengths, the following problems could arise.

PRACTICAL NOTE **Underdetermined** The data could be such that some edges are left undetermined. In these cases we cannot do anything for these edges. In particular there will never be enough information to place the root. In terms of the above example, we will be able to determine $e + f$, but neither e nor f by themselves. We place the root arbitrarily at the midpoint of the two last subtrees joined together. For a LS formulation this known underdeterminacy should be avoided and we should use a single variable $x = e + f$ instead of e and f.

Incomplete data If the sequences are too distant (for example A and C in Figure 8.19), the alignment may be of very poor quality and it is better to discard the particular distance estimate (make its variance $\sigma^2 = \infty$ is enough). In this case, the system may also become underdetermined or the tree may become disconnected. Computationally, to prevent numerical problems, we add to the sum of squares the sum of all the squares of the edge lengths multiplied by a very small factor (e.g. 10^{-10}). Two cases are possible.

Case 1: if the system is fully determined, the extra term has virtually no effect.

Case 2: if the system is initially underdetermined, the extra condition will make it fully determined and force the undetermined lengths to be zero.

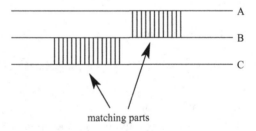

matching parts

Figure 8.19   Alignment problem due to missing or short overlap.

Negative edge length In general, the least squares solution could give edges of negative length. This is not allowed in our model since edges represent evolution, and evolution has a unique (positive) direction. When the least squares solution contains negative edges, we fix these edges to a very small positive length, say 0.01 PAM, and recompute the least squares on the remaining edges. It should be mentioned that having negative edges is a clear indication that the data are not fitting a tree model, i.e. poor data. This could be due to the wrong topology, to very short sequences being aligned or very distant species being compared.

Problems with distances One may wonder why some distances may be undefined if all the others are defined. In sequence alignment this is always possible, and it is also possible that the triangular inequality does not hold. The examples that cause problems are shown in Figure 8.19. Here A and B have a short but good alignment and so do B and C. But there is no common ancestry between A and C. In such cases the distance between A and C could be undefined or completely wrong while the other distances are well defined.

8.4.4 Improving a tree: swapping heuristics

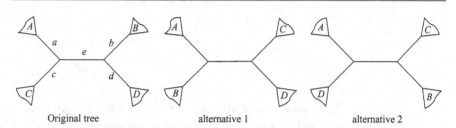

| Original tree | alternative 1 | alternative 2 |

Figure 8.20   The three alternative quartets.

Once a tree is built, and we have distances from the least squares fitting, we look for alternative topologies which may have a better fit, i.e. a smaller error. This is the iterative part of the CHE procedure. We would like to explore neighboring topologies or topologies which are similar to the one we have. This is motivated by two arguments.

(i) Saving computation: if we recompute the tree after a small change we expect to reuse previous results.
(ii) If our tree is a good tree, we should search for better trees "nearby." In the following algorithm we define which are the trees "near" a given tree.

4-optim (nearest neighbor exchange or neighbor swapping)

4-optim inspects all the alternative permutations that can be done for each internal edge, i.e. an edge which is not attached to a leaf. An internal edge defines a quartet: a particular topology which relates four subtrees. 4-optim explores all possible alternative quartets (see Figure 8.20).

4-optim algorithm Input a tree with an internal edge e, where A, B, C, D represent subtrees, all in planar form.

```
curr_error := edge_error(tree);
improved := true;
while improved do
```

```
improved := false;
for e in [internal_edges_not_attached_to_leaves] do
    for t in [alternative_topologies(e)] do
        new_error := edge_error(t);
        if new_error < curr_error then
            improved := true;
            curr_error := new_error;
            tree := t
        fi
    od
od
od;
```

To compare the alternatives we do not need to recompute all the edges of the tree, only the five edges involved in the alternatives. That is e and the four edges a, b, c, d at its ends. For each alternative, selected or not, only five edges need to be computed.

4-optim is guaranteed to terminate, as each time we select a different tree we decrease the norm of the errors. So our search strategy is to look for an alternative topology which has a lower error norm. There are $n - 3$ internal edges and two alternative topologies (neighbors) per internal edge, which means $2(n - 3)$ neighbor trees for every tree with n leaves. This is now the same scenario that we studied for best basis in Section 4.1 and we can apply the techniques described there. In particular we can do a steepest descent or early abort searching strategy. The code, as written above, is an early abort strategy.

5-optim

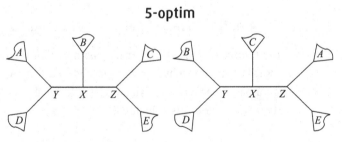

Figure 8.21   **Two alternative configurations in 5-optim.**

Alternatively, instead of looking at each suitable edge, we could look at each suitable internal node with a configuration of five subtrees, as shown in Figure 8.21.

The internal node X which has at least two adjacent internal nodes Y and Z generates 15 possible topologies, our original and 14 alternatives. As for 4-optim we can perform a local least squares fit over seven edges to obtain the optimal length of the edges, and select the best topology.

Of the 14 alternative topologies, four would have been explored by 4-optim on the two edges Y–X and X–Z. Consequently for each internal node and configuration we explore ten new neighboring trees with 5-optim. It can be applied after 4-optim or concurrently with it.

Both 4-optim and 5-optim are ways of finding neighboring trees to explore in our search for a better topology. These are not the only choices, it is possible to find other reasonable sets of neighbors – see the online exercise.

8.5 Phylogenetic trees based on character information

A character is a heritable trait possessed by an organism. Characters can be derived from observable properties or can be derived from molecular sequences (amino acids of a protein or bases of DNA). Characters are usually described in terms of their states, for example:

Character	States			
Skin coverage	hair	feathers	scales	none
Eye color	blue	green	brown	black
Vertebrate	yes	no		
Mammal	yes	no		
DNA at position x	A	C	G	T
AA at position x	R	N	S	C ...

For DNA and amino acids, a multiple sequence alignment (MSA) is normally computed between homologous sequences and each position (column of the MSA) is considered a character.

It is always assumed that a character has a finite number of states and that transitions between these states are equally difficult (no transition is "preferred"). Characters having only two states (e.g. true and false) are called binary characters.

Characters can be informative or uninformative. A character which has the same state for all species is uninformative. Similarly a character which has different states for all species is also uninformative. For example:

Species	Characters						
	c_1	c_2	c_3	c_4	c_5	c_6	c_7
Lettuce	C	N	G	T	K	S	N
Rice	C	N	G	T	T	Q	T
Mosquito	M	N	G	D	K	S	G
Monkey	M	N	G	R	K	N	R
Human	M	N	G	R	K	S	S
	I	U	U	I	I	I	U

where I is informative and U is uninformative. Uninformative characters can be ignored, they do not contribute to the tree and cost only computation time and space.

The goal is to produce trees which explain evolution with a minimum number of changes. Minimum number of changes is synonymous with parsimony. This is quite different from distance methods where every pair of individuals/sequences is reduced to a single number, its distance.

8.5.1 Parsimony

Parsimony selects the tree that minimizes the number of character changes needed to explain the evolution of all leaves. This minimum number of changes is the optimization criterion. This is based on *Occam's Razor*, which states that the simplest explanation of an event is usually the right one. In this case "simplest" is interpreted as "fewest changes."

As with the distance based methods, given a topology it is relatively easy to compute the minimum number of character changes required. The difficult problem is to find the topology which will give the minimum number of character changes.

Computing the minimal number of changes

Given a tree and the characters of the species at the leaves, we can compute the characters at each internal node that minimize the number of changes. The characters at the internal nodes are either characters or sets of characters. A set indicates that all the choices will have the same cost, i.e. the same minimum number of changes. This is a very simple form of dynamic programming – the algorithm proceeds bottom-up: for every character at each internal node we intersect the two sets of possibilities. If the intersection is empty we take the union and we count $+1$ changes.

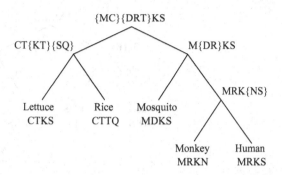

Figure 8.22   Character matching of lettuce, rice, mosquito, monkey and human with six changes.

Using the informative characters c_1, c_4, c_5 and c_6 of the previous example we get the tree in Figure 8.22, which requires six changes.

Some explanations:

- The parent of monkey and human is assigned MRK{NS}. The first three characters coincide and hence the intersection is not empty, and that is what we keep. The last character is different, the intersection is empty, and so we keep the union {NS} and add 1 to the total number of required changes.
- The parent of mosquito–monkey–human has non-empty intersections for the first, third and fourth characters. Notice that the fourth character is now just S (the intersection of S and {NS}). The second character is the union {DR} and one new change has to be counted for.

As usual with dynamic programming the forward computation (in this case bottom-up) gives us the optimal (minimal) number of changes. We could then assign the characters at the internal nodes in the backtracking (top-down) second phase. Every set at the root represents a possible choice of characters contained in the set. For example, we could choose the characters MDKS at the root. Once this is done, the internal nodes will be as shown in Figure 8.23.

Other selections for the root (or anytime we have a set in the backwards phase) will produce trees which have the same cost (number of changes required to explain the leaves), see Figure 8.24.

8.5.2 General parsimony

Parsimony as described above counts the plain number of changes. This may be too crude in some cases, as mentioned before.

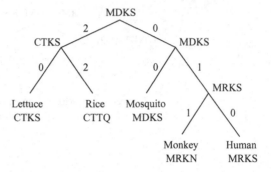

A tree for character matching of lettuce, rice, mosquito, monkey and human. The edge labels indicate the number of character changes for that branch.

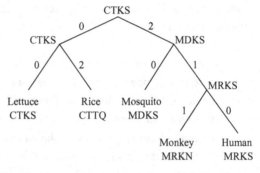

Another possible tree for character matching of lettuce, rice, mosquito, monkey and human, also requiring six changes.

(i) Acquiring a character may be much more difficult than losing it. (This is particularly applicable to binary characters.) In general there could be different costs for character transitions. Another example would be amino acids taken as characters. Some pairs of amino acids are similar (low cost transition), some pairs are very different (high cost transition).

(ii) Some characters have more importance than others and should be weighted differently, for example vertebrate versus eye color.

These and other exceptions are handled by using full dynamic programming to evaluate the number of changes for a given topology.

Example The characters Chlorophyll, Vertebrate and Color of the species lettuce, rice, mosquito, monkey and human. The transition costs are given by:

Chlorophyll		
from	to	
	true	false
true	0	1
false	3	0

Vertebrate		
from	to	
	true	false
true	0	4
false	4	0

Color			
from	to		
	green	brown	pink
green	0	1	1
brown	1	0	$\frac{1}{2}$
pink	1	$\frac{1}{2}$	0

The costs for the tree are calculated with the following algorithm:

```
for each internal node i do #(bottom up)
   for each character c do
      for each state s of c do
         cost[i,c,s]:= sum( min( transit[c,s -> t] + cost[d,c,t]) ),
      od;
   od;
od;
```

Explanations The `min` is computed over all states `t`. The sum is computed for all descendants `d` of `i`.

`cost[i,c,s]` represents the tables of optimal costs at each internal node `i` for character `c` having the state `s`.

`transit[c,s->t]` represents the cost of character `c` going from state `s` to state `t` where typically the cost of not changing, `transit(c,s->s)`, is zero.

The above loop represents the forward phase of the dynamic programming. Once this is done we can, if desired, compute the internal node assignments. At the root we select for each character the state with the lowest cost.

The result of these calculations is the tree in Figure 8.25, next page, where the cost vectors in the tree mean the following:

Chlorophyll	Vertebrate	Color
has	is	green
—	—	brown
		pink

Let us calculate some entries of the cost matrix `cost[i,c,s]` for the root of the animals (mosquito, monkey and human).

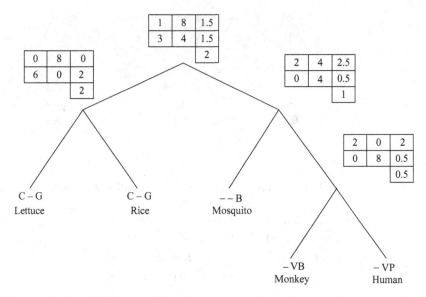

Figure 8.25 The tree calculated with general parsimony.

Has no chlorophyll The mosquito has no chlorophyll, so in its cost matrix this entry would be zero, in the cost matrix of the root of primates there is a zero for no chlorophyll too $\Longrightarrow 0 + 0 = 0$, since nothing is changed.

Has chlorophyll The mosquito has no chlorophyll, and there is only one way to reach this from having chlorophyll – by losing it. So the cost is $0 + 1$ (for the change), the same holds for the root of primates, and the total sum is 2. Alternatively you could set the cost matrix entry for "has chlorophyll" for mosquito to infinity and just use the above algorithm.

Is vertebrate This gives a cost of zero for the root of the primates, and a cost of four for changing from vertebrate to not vertebrate for mosquito, so the sum is four.

Green color For mosquito the cost is one (for changing from green to brown), for the root of the primates we have to take the minimum of three costs: 2 for staying green, $0.5 + 1$ for changing from green to brown, $0.5 + 1$ for changing from green to pink, so the minimum is 1.5. Summed with the cost for mosquito we get 2.5.

We suggest you do the rest of the calculations as an exercise, so that you understand the algorithm completely. In our example the minimum costs at the root are $1, 4, 1\frac{1}{2}$ giving an optimal cost for the tree of $6\frac{1}{2}$. In the example there is some ambiguity because of equal costs. We choose arbitrarily (all choices result in the same cost) C–G for the root. The root of both plants is assigned the same C–G. The root of the animals is assigned –B and the root of the primates –VB.

Notice that even though we work with all the characters together, each character is completely independent of the others. Every topology has a cost, and hence we can use the CHE algorithms (neighbor joining, 4-optim, 5-optim) described earlier in Section 8.4.4, to find initial topologies and then improve them. In the simplest case of parsimony (all changes having the same weight) or if the transition cost matrices are symmetric, the selection of where the root is placed is arbitrary and does not affect the minimal number of changes required for the tree. So the root could be positioned on any edge. Hence the resulting optimal tree is unrooted.

If the transition cost matrix is not symmetric, then the position of the root influences the cost of the tree. We cannot work with unrooted trees any longer, we have to work with rooted trees in this case. The computation is more expensive but we obtain a rooted tree as a result.

See the interactive exercises "Phylogenetic trees" and "Parsimony."

8.6 Special cases allowing easy tree construction

8.6.1 Strict character compatibility

Strict character compatibility, also called *perfect phylogeny* requires that every character change is done at most once. The reasoning behind this requirement is that some character changes are so peculiar that in nature they can only happen once.

For example, the character "being a mammal" implies that all mammals have a common ancestor, a widely accepted view in biology. The logic is that such a sophisticated characteristic cannot arise twice in evolution, i.e. mammality was not invented twice. Note: this requirement is too demanding for nature in general; there are examples of features which were invented multiple times, for example "having wings" (insects, dinosaurs and birds, bats). Here we will consider the case of binary characters only. The algorithm for building a tree with strict character compatibility goes through all the characters and produces an edge (splits the set of species in two) for every character/value.

Perfect phylogeny (strict character compatibility) – an example

Our goal is to construct a perfect phylogenetic tree. Its leaves correspond to the species, and its inner nodes correspond to sets of species which share a number of characters. For each character shared by more than one species, there must be an associated node in the tree which is an ancestor of all species which possess this character. We will label these internal nodes with the character name.

Example 1 Let us consider the four species man, dog, cow and pig. Suppose each species has some of the characters numbered 1 through 6, as shown in the following table:

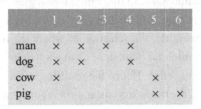

	1	2	3	4	5	6
man	×	×	×	×		
dog	×	×		×		
cow	×				×	
pig					×	×

With this combination of species and characters, tree construction is possible. Not all combinations of characters allow us to build a perfect phylogeny. As a test we could run the algorithm and see whether it fails. If a set of characters does not allow a perfect phylogeny, then there exists a pair of characters for which the four possible combinations (true, true), (true, false), (false, true), (false, false) appear. So if pig had additionally character 3, a perfect phylogeny would be impossible.

Starting at the root The root is associated with the set of all species, i.e. {m, d, c, p}. The algorithm in pseudocode is:

```
root_of_tree := set_of_all_species;
for each character do
    node := root_of_tree
    if all_same_character(node) then
        next
    fi;
    while node is_not_a_leaf do
        if all_same_character(left(node)) then
            node := right(node)
            if all_same_character(right(node)) then
                node[label]:=character;
                node:=right(node);
                break
            fi;
        elif all_same_character(right(node)) then
            node := left(node)
        else error("no perfect phylogeny with these characters")
        fi
    od;
    if not all_same_character(node) then
```

```
                split node into two sets according to character;
                node[label] := character;
                replace node with a new subtree with two new nodes
          fi
     od;
```

The execution of the above code with our example will produce the steps shown in Figures 8.26–8.29.

Figure 8.26 The root of the tree.

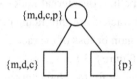

Figure 8.27 The tree after processing the first character.

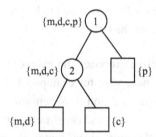

Figure 8.28 The tree after processing the second character.

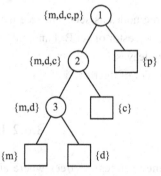

Figure 8.29 The tree after processing the third character.

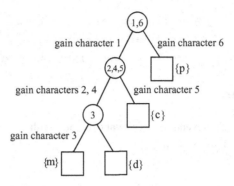

Figure 8.30   The complete tree.

The rest of the characters will not cause any further changes except for the additional labelling of internal nodes with characters. (The labels in the nodes indicate which characters are split in the next branching step.) We can now label the branches with the gain or loss of characters. Each character will be gained or lost in only one branch (Figure 8.30).

Notice that for the characters at the root we can assume that they are either gained on one side or lost on the other. We normally prefer to think of characters gained, that is why we placed "1 gained" to the left and "6 gained" to the right. Alternatively, 1 and 6 could be assumed to be present at the root and 1 is lost on the right whereas 6 is lost on the left.

Interactive exercise The matrix at the top indicates which species (m, d, c, p) has which characters: for example, 1, 2, 4 in the *m*-row means that species *m* has the characters 1, 2 and 4. When you press the "Start calculation" button, the applet checks the input and constructs a phylogenetic tree, if possible.

The "Default values" button re-inserts the input data displayed at start-up.

Output is in two forms

(i) Tree nodes as sets. The tree is represented by a set of sets of species. Set A is an ancestor of set B if and only if B is a subset of A. Sets with only one element are leaves.

(ii) A plot of the tree.

8.6.2 Ultrametric trees

Ultrametric trees are trees where all the leaves are at the same distance from the root. This is usually associated with the notion of an evolutionary clock, where all the species evolve at the same rate. While this is very appealing, estimating accurate times of evolution is very difficult and hence this method is seldom useful. If the data satisfy the ultrametricity condition, the UPGMA algorithm (see Section 8.4.1) will produce the correct tree.

8.6.3 Additive trees

Additive trees represent *exactly* a set of distances between species. Distances between species are represented by a symmetric matrix. The distance between two leaves on the tree is the sum of all the branch lengths in the path from one leaf to the other. Notice that in general, given a set of $n(n-1)/2$ distances it is not possible to find the $2n-3$ internal edges that will satisfy *exactly* all the distances. The neighbor joining algorithm[14] will produce a correct tree for additive trees. When the distances do not satisfy the additivity property exactly, we must find the tree which produces the smallest error and this becomes a much more difficult problem.

8.7 Evolutionary algorithms or genetic algorithms

Evolutionary algorithms (EVA) or genetic algorithms (GA) are a class of algorithms which attempt to copy the evolution of living organisms to solve an optimization problem. In this sense, evolutionary algorithms cannot be defined very precisely, they are more a philosophy or an approach to solving optimization problems.

The more we tailor the EVA to our problem, the more efficient it will be. While our description is generic, our EVA will be problem dependent. Completely generic genetic algorithms are seldom useful.

The ideas that we use from the evolution of life forms are:

- selection of the fittest,
- random mutations (drift),
- sex (combination of two individuals/solutions), and
- a population that evolves in parallel.

These ideas are quite powerful (and effective if we look at the wonderful forms of life that they have created) and quite appealing to computer scientists. Their success at solving problems is mixed – some successes and some failures.

---

[14] M. Aitou and N. Nei, The neighbour-joining method: a new method of reconstructing phylogenetic trees, *Molecular Biology and Evolution*, **4** (1987) p. 425ff.

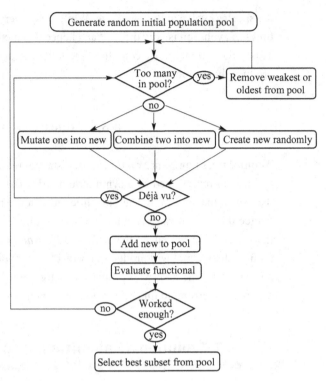

Figure 8.31　The general scheme of an evolutionary algorithm.

We will first define the characteristics of problems that are successfully solvable by evolutionary algorithms.

(i) The problem must be an optimization problem, i.e. some value or values are to be maximized. We will call "functional" the function or program which computes the values to be optimized.

(ii) The functional is a complicated function, usually not having a closed mathematical expression.

(iii) The solution depends on many parameters.

It should be noted that EVAs are not exclusively computer algorithms, they could be purely physical constructions or a mixture. Often the evaluation of the functional is the (computer) simulation of a real physical system. For the rest of this section we will use tree building as an example. The general scheme of an EVA is described in Figure 8.31.

EVAs try to mimic natural evolution. In this sense, the term *population pool* stands for a set of solutions to our problem. A member of the population pool is identified by a set of parameters and other auxiliary information (in particular functional value and age). In our example of a tree, an individual will be represented by an unrooted tree. A good representation is one which allows easy generation of correct individuals, easy evaluation, combination and detection of identical members.

First we explain the boxes in Figure 8.31 in more detail.

Remove weakest This stands for removing the solution which has the worst functional. It may also apply to removing the solution which is the oldest in the set (if it is not the best), or a combination of both criteria. Quite often we add randomness to this step, e.g. remove at random one of the 10% worst solutions.

Too many? We usually keep a sizeable population, say between 100 and 1000 solutions. This number may depend on three things.
- How hard is the problem? For harder problems it is better to have larger populations.
- The computation time. The higher the computation time is for generating new individuals and evaluating their functional, the smaller the population size.
- The expected number of iterations to reach optimality.

The choices downward of "Too many?" are taken randomly with some probabilities which reflect how expensive it is to compute a new solution (in each of the three options) and how likely they are to be better.

Mutate one into new Change an existing solution in a random way. This random mutation should not generate an invalid solution, so for example a rotation/swapping in a tree (see Figures 8.32 and 8.33) is allowed. Inserting repeated leaves is not allowed. The amount of mutation should also be tuned. At the beginning of the optimization, when we are possibly far from the optimal solution, the numbers of random mutations could/should be high. When we are closer to the optimum, the number of mutations should be lower.

Combine two into new Here we take two solutions and combine them into one. Again the resulting solution should be valid. In the case of trees this means finding a branch in each tree which splits the trees in the same leaf-sets. Then we recombine the crossovers.

Create new randomly Create a random, valid solution with the same procedure that is used for the initialization. If the individual is represented by a set of parameters, generation of a new individual is equivalent to generating a consistent set of random parameters. Particular ranges, distributions and interrelations will improve the quality and rate of consistent new individuals.

Déjà vu? This step is useful when solutions are likely to reappear. Of course we do not need to duplicate any solution, so if a solution reappears it can be safely discarded.

What to avoid

Some authors become too literal about life evolution and construct genomes, conversions from the genome to a solution, mutate randomly the genomes. These are usually termed *genetic algorithms*. While GAs are more general, they are usually less efficient and less effective.

EVAs are very general tools. If a specific algorithm has been developed for a problem, it is very likely to be more effective. Purely numerical problems are usually better solved by the wealth of knowledge accumulated in numerical optimization (constrained or unconstrained).

When to use EVAs

- For a new problem whose structure and complexity are unknown. Once algorithms specific for the problem have been developed, it is rarely the case that an EVA will be competitive.
- For ill defined or poorly defined problems, i.e. when the parameter space is complex and possibly difficult to define.
- Parallelism is a big plus for EVAs. In the current cost/effectiveness market it is very appealing to have algorithms that can be effectively parallelized, as n slow machines are much cheaper than one n times faster machine (which may not even exist). Parallelism can be introduced in EVAs at three levels.

  Lowest level: the creation of new members of the population from an existing pool can be done in parallel.

  Medium level: various populations can be grown in parallel, and every so often they can be randomly mixed or repartitioned. (This mimics quite well living species that are separated for geographical reasons but every so often have the chance to mix again.)

  Highest level: various independent populations can be grown independently of each other only to be joined/selected at the end.

The reader should notice the parallelism between EVAs and CHE procedures. CHE optimization can be viewed as a specialization of EVAs.

Examples of mutations

Generation See Section 8.4.1 on UPGMA.

Mutation 1 Rotation (at a random node), see Figure 8.32.

Mutation 2 Swapping (between two random nodes) see Figure 8.33. This is equivalent to 4-optim. Other examples of mutations are 5-optim.

Crossover events A *crossover event* occurs between solutions (individuals), see Figure 8.34. The two solutions are combined into one (or two) new ones (like

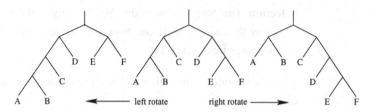

Figure 8.32   Mutation 1, rotation (change position of the root).

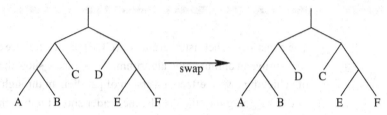

Figure 8.33   Mutation 2, swapping (nearest neighbor interchange).

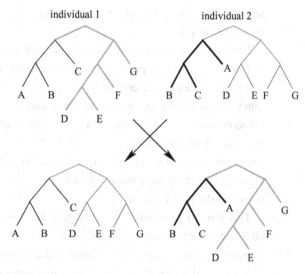

Figure 8.34   Mutation 3, crossover.

a father and a mother combining their genes to produce an offspring). In our example, this corresponds to swapping subtrees between the two individuals. In this case the parts should have at least three leaves, otherwise the crossover has no consequence. The *crossover rate* is the average number of crossover events two individuals undergo in a generation.

Selection The *fitness* of an individual is determined by computing the functional over the individual. In our case, for character trees the number of changes; for distance trees $\|\mathbf{D} - \mathbf{T}\|$.

The relative rates of mutation and recombination may be fixed or changed dynamically.

8.8 Randomness, parallelism and CHE procedures

The constructor–heuristic–evaluator (CHE) procedures are a very convenient way of structuring an optimization algorithm. We will now show that they also provide a way of achieving a very effective version of parallelism through randomization.

Before going into the details, the reader should notice the similarity between the EA/SD algorithms described for best basis (see Sections 4.3.1 and 4.3.2) and the CHE procedure (Section 8.2.2). Basically, the neighbors in EA/SD correspond to the heuristic search of the CHE procedure. The constructor in EA/SD is just a totally random solution. So most of the comments that we made for CHE are directly applicable to EA/SD.

The key ingredient that we will add is randomness. We will allow the constructor and/or the heuristic to produce results which depend on random choices and hence will likely produce different results when executed more than once. Randomness may imply that some choices are not optimal, and we will accept this possibility. Using randomness, two runs of the same algorithm/problem will explore different parts of the solution space and possibly return different solutions. It is this aspect that is important: the runs are completely independent and most likely different. We run the problem in parallel in as many processors as we can, and when all the runs finish then we select the best solution.

This type of parallelism is called high-level parallelism (or embarrassingly parallel, since each execution is independent) and is easy to implement. The parallel computations do not need to communicate at all with each other, so no synchronization is needed except for termination. Both the constructor and the heuristics may be randomized.

8.8.1 Concluding additional information from different runs

Let us assume for a moment, that we use a constructor which selects a totally random initial solution, i.e. in the space of all solutions one is chosen randomly. Then we apply

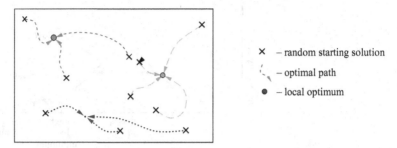

Figure 8.35   Testing the solution space with different random solutions.

the heuristics (randomized or not) and we converge to a local minimum. This can be visualized by selecting random places in a mountainous terrain and then descending as far as we can, like a drop of water would do.

When we run this optimization in parallel many times, the number of different local optima gives us a very good picture of our problem (Figure 8.35). For example, if we run it 1000 times and we obtain in all these runs only three local optima, we have learned that our space is in its majority very well behaved with large "watersheds" that lead to a few (best guess is three) local optima. If instead we run the procedure 1000 times and we get 990 different local optima the space is probably very irregular and the hope of having explored a significant part of it is illusory.

More precisely, if we run n times our optimization procedure and we obtain m different local optima, then the estimate of the total number of minima is:

$$
\begin{cases}
\infty & \text{if } n = m \\
\frac{n^2}{2k} - \frac{(2k+3)n}{6k} + \mathcal{O}(k), \text{ for small } k \text{ and large } n & \text{if } n = m + k \\
x, \text{ where } x \text{ is the positive solution of } x \cdot (1 - e^{-n/x}) = m & \text{for large } n.
\end{cases}
$$

8.8.2 Sensitivity analysis suitable for parallel CHE

The information obtained from several local minima can be used to analyze the confidence that we have in the result. This analysis can be done in any scenario, but the parallel execution with randomness is the preferred one. To do this we will collect all the local minima (position and functional value) that we find in our runs. We will describe our optimization problem as a function of two groups of arguments: $F(X, D)$, where X is the vector of variables on which we are doing the optimization and D the input data to our particular problem. In the case of distance phylogenetic trees, X is a tree with labeled edges (distances), D represents the input data, i.e.

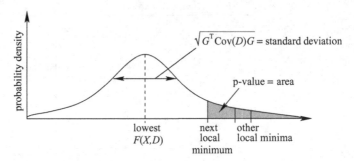

Figure 8.36   **Calculating the significance of the lowest minimum.**

the distances between the leaves and their variances, and $F(X, D)$ represents the functional that we want to optimize, for example minimize.

Depending on the problem we have two ways of attacking it.

Case 1 The function $F(X, D)$ can be differentiated with respect to D. Then if we know (or we can estimate) the errors in the data, we can carry out the sensitivity analysis as follows.

Let $\text{Cov}(D)$ be the covariance matrix of the errors of the data D. (If all the errors are independent of each other, the matrix $\text{Cov}(D)$ is just diagonal and contains the variances of each element of D.)

Let $G = F'_D(X, D)$ be the gradient of $F(X, D)$ with respect to D, i.e. a column vector with the same dimension as D. For any arbitrary point X the total error of $F(X, D)$ due to the errors in the data has variance $G^T\text{Cov}(D)G$. Hence the value of $F(X, D)$, if all the errors in D are normally distributed, is normally distributed with average $F(X, D)$ and variance $G^T\text{Cov}(D)G$ or $N(F(X, D), G^T\text{Cov}(D)G)$. This is true as long as the errors in D are sufficiently small to justify a linear approximation with G.

With this we can now estimate how significant our best local minimum is (Figure 8.36). For this we look where the other local minima are located with respect to the normal distribution of the best $F(X, D)$.

If the p value is very small, we have high confidence that our lowest $F(X, D)$ and hence its corresponding X are the right minimum. If the p value is not too small, the confidence of X being the optimal value decreases correspondingly. In other words, with probability p we could have randomly perturbed data that would give a different minimum, i.e. the minimum is not very stable.

Case 2 The function $F(X, D)$ is not differentiable with respect to the data, or differentiation is too complicated to be realistic. In this case we will run the parallel optimizations with slightly perturbed data. That is to say, not only we will have the

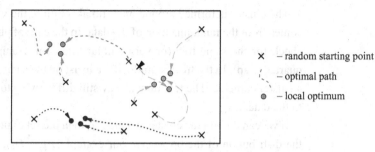

Figure 8.37   Clusters of perturbed minima.

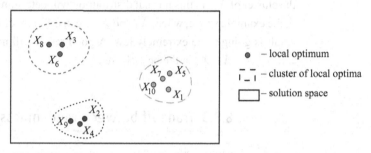

Figure 8.38   Clusters of local optima.

randomness of the initial solution and the heuristic, but we will also modify the input data with suitable random noise. As in case 1, we need to know (or estimate) the magnitude of the errors in the data. In the absence of more precise information, using normally distributed noise with the right variance is the best choice. This is a form of "bootstrapping."

With perturbed data the local minima are very unlikely to coincide: what were simple optima in case 1 will now be clusters (see Figure 8.37). (For a given minimum, the perturbation of the data is going to result in a perturbation of the location of this minimum.) Figure 8.37 shows a simple case, when the displacement of the optimum location is small. It may be that the clusters overlap and that it becomes impossible to see a clear picture.

To resolve the picture we will only keep a small percentage (say 5%) of the best local minima, i.e. the optimizations $F(X_1, D_1) \leq F(X_2, D_2) \leq \cdots$ which give the lowest F values of all our independent runs, D_i being the ith instance of perturbing the original data randomly (Figure 8.38).

If we cannot delimit a cluster around X_1, this is bad news for our confidence analysis, it means that the optimum could be in any place. (An alternative which may resolve this problem is to generate more random points in the hope that the top 5% of the X_i cluster better.)

The cluster is formed around an optimal solution, the variation within the cluster comes from the randomization of the data. In the case of phylogenetic trees we could check for the same tree (or very similar trees, for example differing by only one quartet swap). In the tree example, if we insist on the same tree, then all the points in a cluster coincide. The functionals may still differ owing to the random perturbations to the data.

If we can delimit the cluster around X_1, then the set of points in this cluster defines the distribution of the optima, in our example $F(X, D_1)$, $F(X, D_5)$, $F(X, D_7)$ and $F(X, D_{10})$. We estimate the average and variance of these values and assume a normal distribution. The placement of $F(X, D_2)$ (or the smallest $F(X_i, D_i)$ outside the cluster of X_1) in this normal distribution will determine the p value. It is clear that in the example above, where X_1 and X_2 are in different clusters, the confidence of our result is going to be extremely low. A good result will normally have, for example, X_1, X_2, X_3, X_4, X_5 in the best cluster.

8.8.3 Tradeoff between randomness and quality

Selecting totally random initial solutions may cover the solution space uniformly, but these initial solutions may be too poor to optimize effectively. The random solutions may be too far from the optimal value to ever get there or it may just be too slow to converge. On the other hand, the best solution that we can find with an approximation algorithm is usually deterministic and will not allow effective parallelism.

It would be extremely desirable to "tune" the amount of randomness, having at one extreme a totally random solution and at the other a (good) deterministic solution. If the scale is continuous we can tune this randomness appropriately for each type of problem.

This is possible and relatively easy for greedy approximation algorithms (like UPGMA). Let r be the randomness tuning parameter, where $r = 1$ means a completely random solution and $r = 0$ means a deterministic solution. In a greedy algorithm we make decisions based on some local optimization criteria at each step. (In UPGMA we decide to join the two closest subtrees.) Instead of taking the (locally) best choice we will order the choices according to their optimization criteria: the most desirable choice first, the least desirable last. We will select the ith choice with probability $r^{i-1}(1 - r)$, a geometrical distribution. For example:

choice	probability	UPGMA
1st	$1 - r$	closest pair
2nd	$r(1 - r)$	second closest pair
3rd	$r^2(1 - r)$	third closest pair
$\vdots$	$\vdots$	$\vdots$

If no choice is selected when r is close to one and/or we have few choices, we select a purely random one. The distribution is a geometric distribution which is easily implemented. Notice that $r = 0$ will always give the first choice and hence we will have the standard greedy behavior. For $r = 1$ all choices will fail and we choose a random one which will usually give a completely random solution. r values very close to one will give almost random solutions, r values close to zero will give almost deterministic solutions. In this way we can tune the algorithm continuously between random (and poor) and deterministic (and good).

If there are k choices the following code computes i with the desired distribution, when $U(0, 1)$ is a uniformly distributed random number between 0 and 1 and $U(1, \ldots, k)$ is a random uniform integer between 1 and k.

```
if r=0 then i := 1;
elif r=1 then i := U(1..k);
else i := ceil( ln(U(0,1))/ln(r) );
     if i>k then i := U(1..k) fi
fi;
```

FURTHER READING

L. Gonick, *The Cartoon Guide to Genetics*, Harper Collins, 1991.

D. M. Hillis, C. Moritz and B. K. Mable, *Molecular Systematics*, Sinauer Associates, 1996.

J. Felsenstein, *Inferring Phylogenies*, Sinauer Associates, 2004, a research article on tree construction (going beyond the scope of this book).

Appendix A Methods for function minimization

A1.1 Introduction

BASIC

Function minimization is one of the most important topics in scientific computation. The definition of the problem is quite simple. We are given a function f, called a univariate function if it depends on one parameter or multivariate function if it depends on several parameters. The purpose of the minimization is to find values for the parameter(s) which let f attain its lowest value. Clearly f applied to its parameters must return a real number.

Some definitions relating to function minimization are quite useful:

- univariate – the function to be minimized depends on only one argument,
- multivariate – the function to be minimized depends on many arguments and we want to find the values of all these arguments which give the minimum value,
- global minimum – the lowest value of f for any possible arguments,
- local minimum – a point where f achieves a minimum value for all points near it,
- optimization – usually minimization or maximization of a function which reflects some cost or quality,
- constrained optimization (constrained minimization) – the optimization of a function when the parameters are restricted to some subspace (e.g. only positive values allowed); constrained optimization usually adds inequalities for the parameters which must be satisfied for the solution (e.g. $0 \leq x \leq 10$),
- unconstrained – a minimization problem without constraints, the parameters are allowed to take any values,
- maximization – maximization is completely equivalent to minimization, just replace f by $-f$,
- continuous/differentiable – the function f is continuous/differentiable in its parameters, i.e. $f \in C^0$ or $f \in C^1$,
- real/complex domain – the domains of the parameters are normally the real numbers, very rarely it may be the complex numbers, in which case f should still be a real-valued function.

Here is an overview of the most common methods for function minimization:

univariate methods   analytical method
                        Brent's method (golden section search)
                        parabolic interpolation

multivariate methods   analytical method (including EV decomposition)
                        steepest descent
                        random directions
                        Newton's method
                        conjugate gradients
                        spectral method

A1.2 Univariate methods

A1.2.1 The analytical method of function minimization

BASIC

For continuously differentiable functions $f : \mathbb{R} \to \mathbb{R}$, maxima and minima can only occur where the function has a horizontal tangent. That is, if $f'(x)$ is continuous we find maxima and minima of f where $f'(x) = 0$. If f is twice continuously differentiable, then the sign of f'' decides the curvature at this point, i.e. which kind of extremum we have:

- $f'(x) = 0$ and $f''(x) > 0$ guarantees that there is a minimum;
- $f'(x) = 0$ and $f''(x) < 0$ guarantees that there is a maximum;
- if $f'(x) = 0$, $f''(x) = 0$ and $f'''(x) \neq 0$ then there is an inflection point, i.e. neither minimum nor maximum;
- cases like $f'(x) = 0$, $f''(x) = 0$ and $f'''(x) = 0$ are very rare cases in reality, the analysis is as above exchanging f for f'' and f' for f'''.

From this analysis it is clear that all minima occur where $f'(x) = 0$. If we have a formula for $f'(x)$, it is sometimes possible to solve this exactly and find all candidate values.

Example 1

$$f(x) = \cos\left(1 + x^2\right)$$

If we want to minimize this univariate function we should compute its derivative which is

$$f'(x) = -2x \sin\left(1 + x^2\right).$$

Solving $f'(x) = 0$ we have $x = 0$ or $\sin(1 + x^2) = 0$, which means that $1 + x^2 = n\pi$ for any integer n. In this case we have an infinite number of solutions, two for each positive integer n. These are the two solutions of the quadratic equation $x^2 = n\pi - 1$, i.e. $x = \pm\sqrt{n\pi - 1}$. Since x has to be real, $n > 0$ is a necessary condition for the existence of solutions.

We now try $x = 0$ and see that $f''(0) = -2\sin(1) \approx -1.682\,941\,970 < 0$, so $f(x)$ has a maximum at $x = 0$.

For $x = \pm\sqrt{n\pi - 1}$ with $n > 0$ we get

$$f''(x) = -2\sin(n\pi) - 4(n\pi - 1)\cos(n\pi) = 4(n\pi - 1)(-1)^{n+1},$$

since $\sin(n\pi) = 0$ and $\cos(n\pi) = (-1)^n$.

So for odd n, $f''(x) > 0$ and for even n, $f''(x) < 0$. So f has an infinite number of minima.

The values at the minima (for odd n) are $f(\pm\sqrt{n\pi - 1}) = \cos(n\pi) = -1$, so all these minima are global minima, whereas the minimum at $x = 0$ is only a local minimum.

Example 2

$$f(x) = \frac{x}{1 + x^2}, \quad f'(x) = \frac{(x + 1)(1 - x)}{(x^2 + 1)^2}$$

In this case $f'(x) = 0$ for $x = 1$ and $x = -1$.

The second derivative of f is

$$f''(x) = \frac{2x^3 - 6x}{(x^2 + 1)^3},$$

and $f''(1) = -\frac{1}{2}$ whereas $f''(-1) = \frac{1}{2}$. So f has a minimum for $x = -1$ and a maximum for $x = 1$, i.e. there is only one value of x with a minimum, and accordingly $f(-1) = -\frac{1}{2}$ is the global minimum of the function.

Obviously this method requires the exact solution of an equation, which is not possible in general. Hence we are going to develop methods which compute numerically the minima or the zeros of the derivative.

A1.2.2 Brent's method (golden section search in one dimension)

Finding a root of a function by bisection works on the basis of keeping an interval (a, b), which we know contains a root since $f(a) \cdot f(b) < 0$, and reducing the interval size in every iteration. We take the midpoint $x = (a + b)/2$ and decide which of $f(a) \cdot f(x)$ or $f(x) \cdot f(b)$ is less than zero.

The same idea can be applied to finding minima. In this case we need three points (a, b, c) and the condition becomes $f(a) > f(b)$ and $f(c) > f(b)$. Notice that if $f(x)$ is continuous in (a, c) the above conditions guarantee that there will be at least one local minimum in (a, c). Notice also that $f(b)$ is the minimum functional for the three points, which is our invariant.

To reduce the interval size, we have to choose a point inside (a, c) and determine which are the three points in the new interval that will contain a minimum. Let the new point be x, then if $f(x) < f(b)$, x will be the new midpoint, otherwise b remains the midpoint. In all cases x and b are part of the new triplet, either a or c is kept, and the other removed (see Figure A1.1).

In bisection, the new point x is chosen to be the midpoint between a and b. Since we select either (a, x) or (x, b) as the new interval, selecting the midpoint guarantees that even in the worst case the interval decreases by a factor of $\frac{1}{2}$ in each iteration.

For the minimization problem we have a very similar goal: that the configuration of (a, b, c), i.e. the ratios between the lengths, be maintained and that the total interval decreases as much as possible even in the worst case. This will be discussed in detail in the next section. The table compares bisection and Brent's minimization.

	Bisection	Brent
prerequisites	f continuous in (a, b)	f continuous in (a, c)
goal	find $f(x) = 0$	find min $f(x)$
points used	(a, b)	(a, b, c)
condition	$f(a) \cdot f(b) < 0$	$f(a) > f(b)$ and $f(c) > f(b)$
next point	$x = \frac{a+b}{2}$	$x = a + c - b$
interval ratio	$\frac{1}{2}$	$\phi = \frac{\sqrt{5}-1}{2}$
number of iterations to reach relative error δ	$\log_2\left(\frac{1}{\delta}\right)$	$1.44 \cdot \log_2\left(\frac{1}{\delta}\right)$

Stopping criteria

In bisection it is not uncommon to run the bisection until a and b are separated by the minimum amount possible in their floating point representation, usually called a "ulp" (unit in the last place). The last few iterations may respond to random floating point errors, a consequence of the evaluation of $f(x)$, but this is normally not a serious concern.

In minimization this becomes much more of a concern because of the expected flatness of the function at the minimum. If we expand the function in a Taylor series around its minimum (let us call it b) we obtain:

$$f(x) \approx f(b) + f'(b)(x - b) + \frac{1}{2}f''(b)(x - b)^2. \tag{A.1}$$

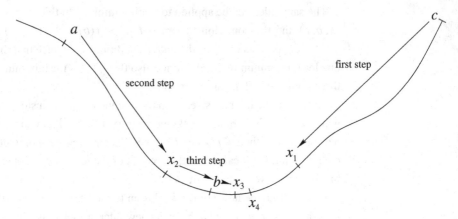

Figure A1.1 An example of successive bracketing of a minimum. Here is an overview of the bracketing points in the first five steps:

step	bracketing points			probe
1	a	b	c	x_1
2	a	b	x_1	x_2
3	x_2	b	x_1	x_3
4	b	x_3	x_1	x_4
5	b	x_3	x_4	

We assume that the function has enough derivatives which are continuous and at the minimum $f'(b) = 0$. It is clear that if the last term, relative to the first term, is smaller than ε,[1] it will not have an effect on the value of $f(x)$ and hence it will not allow us to determine the location of the minimum. This limiting condition is expressed as:

$$\frac{1}{2}\left|f''(b)\right|(x-b)^2 \approx \varepsilon \cdot |f(b)|$$

$$(x-b)^2 \approx 2\varepsilon\left|\frac{f(b)}{f''(b)}\right|$$

or $$\left|\frac{x-b}{b}\right| \approx \sqrt{\varepsilon}\sqrt{\frac{2|f(b)|}{b^2\,|f''(b)|}}. \qquad (A.2)$$

The left-hand side is the relative error of the found minimum x, compared with the absolute minimum b, that we want to approximate. The right-hand side is proportional to $\sqrt{\varepsilon}$, whereas for finding a root of f we have a proportionality to ε. This means that numerical minimization is inherently less precise than root finding or other procedures. Bisection achieves a relative error of ε (which means roughly $\log_2(1/\varepsilon)$ bits of precision) while minimization achieves a relative error of $\mathcal{O}(\sqrt{\varepsilon})$ which means $\log_2(1/\sqrt{\varepsilon}) = \frac{1}{2}\log_2(1/\varepsilon)$. Compared to bisection, minimization obtains a result which has half of the bits of precision and costs about 72% of the number of iterations.

---

[1] The machine epsilon is defined as the smallest value, such that float $(1 + \varepsilon) > 1$.

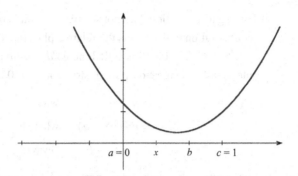

Selecting the next point (see also Figure A1.1).

Selecting the next point

In bisection the next abscissa to inspect is the midpoint. For minimum finding we have two choices, on which side to select the point – in interval (a, b) or in interval (b, c) and at which relative position (Figure A1.2). The optimal selection is the one which preserves the relative lengths of the intervals (in any possible outcome) and hence can be applied repeatedly.

We are going to work in relative terms, so we can assume that $a = 0$ and $c = 1$ and b is some point in between. We will also assume that $b > \frac{1}{2}$, since it is clear that $b = \frac{1}{2}$ cannot be correct for our procedure. (At the next step it would leave intervals of unequal size.) Since the problem is symmetric, we have the liberty to rename (a, b, c) as (c, b, a) should this be necessary.

The new point should clearly be selected in the largest interval, otherwise there is no possibility of preserving the ratios, since the smallest interval could get even smaller. So Figure A1.2 is an accurate representation of where we should choose x for $(0, b, 1)$. We will call $(0, b)$ the "large" interval and $(b, 1)$ the "small" interval.

Depending on the function, the next interval will be either $(0, x, b)$ or $(x, b, 1)$. It is clear that if $(b, 1)$ was the small interval initially, it will be the large interval in the next step. So from $(x, b, 1)$ we conclude that (x, b) is the small interval of the next step and that $(0, x)$ is equal in length to the next large interval $(b, 1)$, i.e. $x - 0 = 1 - b$.

Finally the ratios between small and large intervals have to be kept between steps:

$$\frac{\text{previous small}}{\text{previous large}} = \frac{\text{next small}}{\text{next large}} \quad \text{or}$$

$$\frac{1 - b}{b} = \frac{b - x}{x} = \frac{2b - 1}{1 - b} \quad (\text{using } x = 1 - b)$$

$$1 - 2b + b^2 = 2b^2 - b$$

$$b^2 + b - 1 = 0$$

$$b = \frac{-1 \pm \sqrt{5}}{2}.$$

The negative solution does not satisfy our initial condition $\frac{1}{2} < b < 1$, so $(\sqrt{5} - 1)/2$ is our solution and gives us the relative placement of b in our interval.

$(\sqrt{5} - 1)/2 \approx 0.618$ is called the *golden ratio* and is usually denoted by ϕ. Computationally when we drop the restrictions $a = 0$ and $b = 1$

$$
\begin{aligned}
x &= a + (c - b) & \text{when } |c - b| < |b - a| \\
&= c - (b - a) & \text{when } |c - b| > |b - a| \text{ or} \\
&= a + c - b & \text{always.}
\end{aligned}
$$

BACKGROUND The supposedly aesthetic properties of the golden mean or golden section were already known to the ancient Greeks. It is widely believed that objects and buildings which have subdivisions proportional to ϕ are pleasing to the eye. On the other hand, sounds with frequencies in the ratio of ϕ are dissonant.

Every iteration reduces the length of the interval by a factor of ϕ (regardless of the choice of (a, x, b) or (x, b, c)). Hence the number of iterations needed to make an initial interval of length $c - a$ smaller than δ is

$$
(c - a)\phi^n \leq \delta \qquad \text{or}
$$

$$
n \geq \log_{1/\phi} \frac{c - a}{\delta} \approx 1.44 \log_2 \frac{c - a}{\delta}.
$$

Convergence is linear, meaning that the exponent of the error decreases linearly (or in terms of bits of precision, each iteration increases the number of bits by a constant amount).

Brent's initialization: exponential search To apply the main iteration we need to find points (a, b, c) which satisfy the ratios and value conditions. To find such points in an efficient manner we use an exponential searching technique. This will give us the points with the right ratios and conditions and will also search for a local minimum efficiently, even when we start our search very far from the minimum or our scale is completely off.

The search will be started from two given points $(a, b) = (a_1, b_1)$ (Figure A1.3). The algorithm will decide in which direction to search and what step sizes to choose. The exponential part of this search means that when we are faced with a function which continuously decreases we increase the size of the successive intervals exponentially. Because the exponents in any floating point system are bounded, we will perform at most $\mathcal{O}(\text{max\_exponent})$ iterations.

Given a and b, assume $f(a) \geq f(b)$. (If this is not satisfied, reverse a and b.) Then we want to find a point c in the direction of a, b such that $(b - a)/(c - b) = \phi$ or $c = ((1 + \phi)b - a)/\phi$. When $f(c) \geq f(b)$ the exponential search terminates, if $f(c) < f(b)$ we replace (a, b) by (b, c) and continue the exponential search.

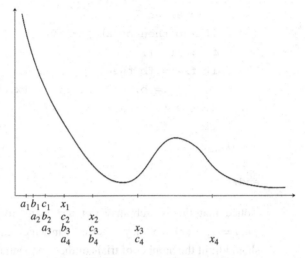

$$a_1 b_1 c_1 \quad x_1$$
$$a_2 b_2 \quad c_2 \quad x_2$$
$$a_3 \quad b_3 \quad c_3 \quad x_3$$
$$a_4 \quad b_4 \quad c_4 \qquad x_4$$

Figure A1.3   **Brent's exponential search – will the minimum be found?**

```
BrentMinimization := proc(f::procedure, a0::numeric,
                          b0::numeric, rel_tolerance::numeric )

# initialization
  a := a0;
  b := b0;
  fa := f(a);
  fb := f(b);
  if fa < fb then
      t := a; a := b; b := t;
      t := fa; fa := fb; fb := t;
  fi;


# exponential search
  phi := ( sqrt(5.0)-1 ) / 2;
  do c := ( (1+phi)*b - a ) / phi;
      fc := f(c);
      if fc >= fb then break fi;
      a := b; fa := fb;
      b := c; fb := fc;
  od;


# minimization
  while abs(c-a) > rel_tolerance*(abs(c)+abs(a)) do
```

```
    x := a+c-b;
    if x=b then break fi;
    fx := f(x);
    if fx <= fb then
          a := b; b := x; fb := fx;
    else c := a; a := x;
    fi;
  od;
  b;
end;
```

Notice that this search may not terminate, for example for functions such as $f(x) = \frac{1}{x}$, $f(x) = x$ or $f(x) = e^{-x}$, in which case a production quality algorithm should limit the number of trials of the exponential search.

A1.2.3 Parabolic interpolation

Brent's minimization method is robust and will converge to a minimum with linear order, i.e. the number of correct digits is proportional to the number of iterations. We already mentioned the similarities between bisection and Brent. So the natural question is: which is the equivalent minimization algorithm to the secant method, which converges with an order higher than linear? The answer is: parabolic interpolation.

	Secant method	Parabolic interpolation
prerequisites	f and f' continuous	f and f' continuous
goal	find $f(x) = 0$	find min $f(x)$
points used	x_n, x_{n-1}	x_n, x_{n-1}, x_{n-2}
approximate f by	a line	a parabola
choose x_n where	line crosses axis	minimum of parabola is
convergence order	$1.618\ldots$	$1.324\ldots$

The formula for the next point, given x_n, x_{n-1} and x_{n-2} is:

$$x_{n+1} = x_n - \frac{1}{2} \frac{(x_n - x_{n-1})^2[f(x_n) - f(x_{n-2})] - (x_n - x_{n-2})^2[f(x_n) - f(x_{n-1})]}{(x_n - x_{n-1})[f(x_n) - f(x_{n-2})] - (x_n - x_{n-2})[f(x_n) - f(x_{n-1})]}.$$

(A.3)

For comparison, the formula for the secant method is

$$x_{n+1} = x_n - \frac{f(x_n)(x_n - x_{n-1})}{f(x_n) - f(x_{n-1})}.$$

In the case of minimization we have an effective check that we can perform for each step: if we are seeking for a minimum, the parabola has to have a positive quadratic term. The quadratic term is given by:

$$a_2 = \frac{f(x_{n-2})(x_n - x_{n-1}) - f(x_{n-1})(x_n - x_{n-2}) + f(x_n)(x_{n-1} - x_{n-2})}{(x_n - x_{n-1})(x_n - x_{n-2})(x_{n-1} - x_{n-2})}.$$

To derive the iteration formula (A.3) we use the following approach:

$$f(x) \approx f(x_{n-1}) \frac{(x - x_n)(x - x_{n-2})}{(x_{n-1} - x_n)(x_{n-1} - x_{n-2})} + f(x_n) \frac{(x - x_{n-2})(x - x_{n-1})}{(x_n - x_{n-2})(x_n - x_{n-1})}$$

$$+ f(x_{n-2}) \frac{(x - x_{n-1})(x - x_n)}{(x_{n-2} - x_{n-1})(x_{n-2} - x_n)}.$$

The idea is that the right-hand side is of degree 2 in x, and is identical to the left-hand side for the three x values x_{n-1}, x_n and x_{n-2}. Since a parabola (with given orientation) is uniquely defined by three points, the left-hand side and the right-hand side describe the same parabola. As we are searching for a minimum of this parabola, let us compute the derivative with respect to x and set it to zero (noting that the denominator does not contain x):

$$f'(x) = f(x_{n-1}) \frac{(x - x_n) + (x - x_{n-2})}{(x_{n-1} - x_n)(x_{n-1} - x_{n-2})} + f(x_n) \frac{(x - x_{n-2}) + (x - x_{n-1})}{(x_n - x_{n-2})(x_n - x_{n-1})}$$

$$+ f(x_{n-2}) \frac{(x - x_{n-1}) + (x - x_n)}{(x_{n-2} - x_{n-1})(x_{n-2} - x_n)} = 0.$$

We multiply with the denominator $(x_{n-1} - x_n)(x_n - x_{n-2})(x_{n-2} - x_{n-1})$ and after some algebraic manipulations we obtain the above result.

A combined approach

We now have two methods, one slow but safe (Brent) and one fast but not as safe (parabolic interpolation). A good combination of both should exploit the best of each at little extra cost. This can be achieved by performing one step of Brent followed by one step of parabolic interpolation. There are several minor concerns to be addressed and these are the subject of an exercise.

A1.3 Multidimensional methods

A1.3.1 The analytical method

Given a function $f(x, y)$ or more generally $f(x_1, \ldots, x_n) = f(\mathbf{x})$, can we use a comparable analytical method like in the univariate case for finding extrema? The answer is yes, and to do so we use the gradient $\nabla f(\mathbf{x}) = f'(\mathbf{x})$ and Hessian $\nabla^2 f(\mathbf{x}) = f''(\mathbf{x})$ which are defined as

$$\nabla f(\mathbf{x}) = f'(\mathbf{x}) = \begin{pmatrix} f_{x_1}(\mathbf{x}) \\ f_{x_2}(\mathbf{x}) \\ \vdots \\ f_{x_n}(\mathbf{x}) \end{pmatrix} = \begin{pmatrix} \frac{\partial}{\partial x_1} \\ \frac{\partial}{\partial x_2} \\ \vdots \\ \frac{\partial}{\partial x_n} \end{pmatrix} f(\mathbf{x})$$

$$= \left(\frac{\partial}{\partial x_1} f(\mathbf{x}), \ldots, \frac{\partial}{\partial x_n} f(\mathbf{x}) \right)^{\mathrm{T}}$$

$$\text{and} \quad \nabla^2 f(\mathbf{x}) = f''(\mathbf{x}) = \begin{bmatrix} \frac{\partial^2}{\partial x_1 \partial x_1} f & \cdots & \frac{\partial^2}{\partial x_1 \partial x_n} f \\ \vdots & \ddots & \vdots \\ \frac{\partial^2}{\partial x_n \partial x_1} f & \cdots & \frac{\partial^2}{\partial x_n \partial x_n} f \end{bmatrix}.$$

Note that the Hessian f'' is symmetric if $f \in C^2$.

In terms of domains:

$$f : \mathbb{R}^n \to \mathbb{R}$$
$$f' : \mathbb{R}^n \to \mathbb{R}^n$$
$$f'' : \mathbb{R}^n \to \mathbb{R}^n \times \mathbb{R}^n.$$

Example For $f(x_1, y_1) = \arctan(x_1) - x_2^2$, we have

$$f'(\mathbf{x}) = \begin{bmatrix} \frac{1}{1+x_1^2} \\ -2x_2 \end{bmatrix} \quad \text{and} \quad f''(\mathbf{x}) = \begin{bmatrix} \frac{-2x_1}{(1+x_1^2)^2} & 0 \\ 0 & -2 \end{bmatrix}.$$

To develop the conditions which have to be satisfied by the gradient and the Hessian for maxima and minima to occur, we examine the Taylor series of f.

For $\mathbf{x} = (x_1, \ldots, x_n)$ and $\Delta \mathbf{x} = (\Delta x_1, \ldots, \Delta x_n)$ the Taylor series of $f : \mathbb{R}^n \to \mathbb{R}$ is given by:

$$f(\mathbf{x} + \Delta \mathbf{x}) = f(\mathbf{x}) + \Delta \mathbf{x} f'(\mathbf{x}) + \frac{1}{2} \Delta \mathbf{x} f''(\mathbf{x}) \Delta \mathbf{x}^{\mathrm{T}} + \mathcal{O}(\|\Delta \mathbf{x}\|^3).$$

By analogy with the univariate case, minima and maxima of f can only occur where f has a horizontal tangent (hyper)-plane, i.e. the necessary condition $f'(\mathbf{x}) = (0, \ldots, 0)^{\mathrm{T}}$ has to be satisfied.

So the Taylor series in an extremal point $\mathbf{x}$ has the form

$$f(\mathbf{x} + \Delta \mathbf{x}) \approx f(\mathbf{x}) + \frac{1}{2} \Delta \mathbf{x} f''(\mathbf{x}) \Delta \mathbf{x}^{\mathrm{T}}. \tag{A.4}$$

Now we need a way to tell whether, at this point where $f'(\mathbf{x}) = 0$, we have a maximum, a minimum or a saddle point. The conditions for the critical points being extrema are given by the signs of the eigenvalues of the Hessian $f''(\mathbf{x})$.

- If all $\lambda_i > 0$ we have a minimum. Such a Hessian is called positive definite.
- If all $\lambda_i < 0$ we have a maximum. Such a matrix is called negative definite.
- If $\lambda_i < 0$ and $\lambda_j > 0$ for at least one i and one j we have a saddle point.
- Otherwise (if at least one $\lambda_j = 0$ and all other $\lambda_i \geq 0$ or all other $\lambda_i \leq 0$) further analyses are necessary, but we will not investigate these rare cases here.

To get these eigenvalues λ, we have to perform an eigenvalue/eigenvector decomposition of the Hessian.

Eigenvalue/eigenvector decomposition

For a symmetric matrix $\mathbf{H} \in \mathbb{R}^{n \times n}$, we have a decomposition

$$\mathbf{H} = \mathbf{U}\mathbf{\Lambda}\mathbf{U}^T \quad \text{with} \quad \mathbf{\Lambda} = \mathrm{diag}(\lambda_1, \ldots, \lambda_n),$$

where the diagonal matrix $\mathbf{\Lambda}$ contains the eigenvalues λ_k and $\mathbf{U} = (\mathbf{u}_1, \ldots, \mathbf{u}_n)$ contains the eigenvectors. If $\mathbf{H}$ is symmetric and real, the eigenvalues λ_k are real and the eigenvectors associated to different eigenvalues are orthogonal, that is $\mathbf{U}^T\mathbf{U} = \mathbf{I}$. The basic properties are therefore

$$\mathbf{H}\mathbf{u}_i = \lambda_i \mathbf{u}_i$$

$$\mathbf{u}_i^T \mathbf{u}_k = \delta_{ik} = \begin{cases} 0 & \text{if } i \neq k \\ 1 & \text{if } i = k. \end{cases}$$

Using the eigenvalue decomposition of the Hessian $f'' = \mathbf{U}\mathbf{\Lambda}\mathbf{U}^T$, we obtain from the Taylor series (A.4)

$$f(\mathbf{x} + \Delta\mathbf{x}) \approx f(\mathbf{x}) + \frac{1}{2} \underbrace{\Delta\mathbf{x}\mathbf{U}}_{\mathbf{z}} \mathbf{\Lambda} \underbrace{\mathbf{U}^T\Delta\mathbf{x}^T}_{\mathbf{z}^T}$$

$$= f(\mathbf{x}) + \frac{1}{2}\mathbf{z}\mathbf{\Lambda}\mathbf{z}^T, \quad \text{where} \quad \mathbf{z} = (z_1, \ldots, z_n) = \Delta\mathbf{x}\mathbf{U}$$

$$= f(\mathbf{x}) + \frac{1}{2}\sum_{i=1}^{n}\lambda_i z_i^2.$$

From this approximation we can derive the above conditions. For example, if all $\lambda_i > 0$ then $f(x_0 + \Delta x) > f(x_0)$ for small Δx, and hence $f(x_0)$ **is** a minimum.

Please see the interactive exercises "Himalaya," "Approximation" and "Approximation 2."

A1.3.2 Steepest descent minimization

Steepest descent minimization is an iterative method to find a minimum, which at each point chooses the direction of steepest descent to approach the minimum. It moves as far as possible in this direction, and then chooses a new direction.

Let $f(\mathbf{x})$ be a function from $\mathbb{R}^n \to \mathbb{R}$, and $f'(\mathbf{x}) = \nabla f(\mathbf{x})$ its gradient. For a vector $\mathbf{\Delta x} = (\Delta x_1, \Delta x_2, \ldots, \Delta x_n)$ of small norm,

$$f(\mathbf{x} + \mathbf{\Delta x}) - f(\mathbf{x}) = \mathbf{\Delta x} \cdot f'(\mathbf{x}) + \mathcal{O}\left(\|\mathbf{x}\|^2\right)$$

(this is just the rewritten Taylor series).

Ignoring the term of second order and assuming that $\mathbf{\Delta x}$ is of constant norm, $\|\mathbf{\Delta x}\| = \varepsilon$, we want to choose $\mathbf{\Delta x}$ so that the right-hand side of the above equation is minimal. $f'(\mathbf{x})$ is a given vector and $\mathbf{\Delta x}$ has bounded norm, so the maximum absolute value of the scalar product $\mathbf{\Delta x} \cdot \nabla f(\mathbf{x})$ arises, when $\mathbf{\Delta x}$ is parallel to $\nabla f(\mathbf{x})$. The minimum value will arise when they are parallel, but have opposite orientation, i.e. when $\mathbf{\Delta x} = -\varepsilon f'(x)/\|f'(x)\|$.

We normally use Brent's search in the direction of the steepest descent, as it is very difficult to estimate how far we can move in the steepest descent direction. Often we do just enough Brent steps to get close to the minimum, not necessarily to full precision, then we compute a new gradient and repeat.

Brent used in a given direction Brent's method is a univariate minimization method. For multidimensional problems we can use Brent's method to search in a particular direction.

Let $\mathbf{d} := (d_1, d_2, \ldots, d_n)$ with $\|\mathbf{d}\| = 1$ be a direction vector. We then minimize the function $f(\mathbf{x} + h \cdot \mathbf{d}) = f(x_1 + hd_1, x_2 + hd_2, \ldots, x_n + hd_n) =: f(h)$ by regarding it as a univariate function of h. The resulting h from Brent's algorithm gives us the best step size to be taken in the direction of $\mathbf{d}$.

Please see the interactive exercise "Minimization."

A1.3.3 Minimization using random directions

The main idea of using random directions is that when we are away from a minimum most directions from an arbitrary point will not be horizontal and hence will decrease to one side. The algorithm consists of repeating the following two steps:

- generate a random direction (in n dimensions),
- apply Brent's minimization in this direction until we are sufficiently close to the minimum.

Random directions is the most robust algorithm for minimization – it is most likely to succeed with any function. Its main advantage is that it does not use derivatives. This is essential if derivatives do not exist, or are very difficult to compute. Overall, it may be more economical, since only f is evaluated. Its main disadvantage is the slow convergence, in particular close to the minimum.

```
d := array(1..n);
fh := h   -> f(x+h*d);
h0 := 1;
to MaxIterations do
    #generate random direction
    for i to n do d[i] := RandomNormal() od;
    h1 := BrentMinimization(fh, 0, h0, TOL);
    if h1<>0 then # move to new point
        for i to n do x[i] := x[i]+h*d[i] od:
        h0 := h;
    fi
od
# x contains the approximation to the minimum;
```

Adaptive parameters The value h_0 in the above algorithm is called an adaptive parameter and it deserves some explanation. To start Brent, we use two points: the first, $h = 0$, implies that we use x, our previous best minimum. The second point is the initial factor in the given direction. Any value would work. However, if this initial value is completely out of scale (for example we use 1 and the minimum happens at $h = 10^{12}$), Brent will make a lot of iterations in its exponential search part, just to reach the right order of magnitude. Nothing is better than to use a previously successful increment. In this way, for a problem on a very peculiar scale only the first iteration will waste time adjusting to the scale, the rest will start sufficiently close.

An adaptive parameter is a parameter used by our functions which is difficult to guess correctly for every problem. We proceed by computing the value and "adapting" it for the next iteration of the algorithm.

Generation of random directions There is a well understood method for generating random directions, but we will explore other, less efficient methods as well, to understand this problem in more detail.

Random directions are equivalent to uniformly distributed points:

in two dimensions, on the circumference of a circle,
in three dimensions, on the surface of a sphere,
in n dimensions, on the surface of a hypersphere.

Construction of such random points, provided that we have a (good) random number generator $\mathcal{U}(0, 1)$ can be done as follows.

Two dimensions Ad hoc $(\cos\phi, \sin\phi)$ with ϕ in $\mathcal{U}(0, 2\pi)$.

Three dimensions If we use a random uniform longitude $(0, 2\pi)$ and a random uniform latitude $(-\pi/2, \pi/2)$, we do not get equally distributed directions. (We get higher densities of points at the poles than at the equator.) Fix for this problem: we use points $\mathbf{d} = (d_1, d_2, d_3)$ with $d_i \in \mathcal{U}(-1, 1)$, i.e. equally distributed points in the volume of a unit cube. We discard all points $\mathbf{d}$ outside the unit sphere to insure a uniform distribution of points inside the unit sphere, and we project these points onto the surface of the unit sphere, i.e. we use the following algorithm for $n = 3$:

```
d := array(1..n);
do norm := 0;
    for i to n do
        d[i] := U(0,1);
        norm := norm + d[i]^2
    od;
    if norm <= 1 then break fi
od;
norm := sqrt(norm);
for i to n do d[i] := d[i]/norm od;
```

n-Dimensions We simply generalize the above algorithm to n dimensions.

Note In the two-dimensional case this is equivalent to selecting a point in the unit square, but rejecting it if it falls outside the unit circle. It is clear that a single trial will succeed with probability $\pi/4$ (the area of the circle divided by the area of the square).

To evaluate the complexity of the method (i.e. how much time it will require to compute) we need to study how many times we will loop before finding a suitable direction $\mathbf{d}$. If we succeed in each run with probability p, then we will run the loop in total

one time with probability p,
two times with probability $(1 - p)\, p$
three times with probability $(1 - p)^2\, p$
. . .
n times with probability $(1 - p)^n\, p$.

The expected (average) number of runs is then

$$\sum_{i=1}^{\infty} i\,(1 - p)^{i-1}\, p = \frac{1}{p}.$$

For the two-dimensional case this is $4/\pi \approx 1.273\,239\ldots$, for the three-dimensional case this is $8/(\frac{4}{3}\pi) = 6/\pi = 1.909\,859\ldots$.

So this simple generalization is not a good algorithm for all n, or is it? Let us have a closer look!

The volume of a hypersphere in n dimensions and its consequences for the generation of random directions What is the probability of generating a point inside the hypersphere, when generating a random point in the hypercube? This is simply the volume of the hypersphere divided by the volume of the hypercube:

Dimension	Corners	$V_{\text{hypersphere}}$	$V_{\text{hypercube}}$	Ratio
1	2	$V_1 = 2$	2	1
2	4	$V_2 = \pi$	4	$\pi/4$
3	8	$V_3 = 4\pi/3$	8	$\pi/6$
n	2^n	?	2^n	?

Now we develop the formula for $V_{\text{hypersphere}}$ for n dimensions. n-dimensional hyperspheres with radius r are defined by coordinates $(x_1, x_2, \ldots, x_n)$ which satisfy the inequality $x_1^2 + x_2^2 + \cdots + x_n^2 \le r$. Let $V_n(r)$ be the volume of the n-dimensional hypersphere with radius r and let us define $V_n := V_n(1)$, then we have $V_n(r) = V_n \cdot r^n$. This observation allows us to determine a recursive formula for V_n by induction.

We choose an arbitrary variable, for example x_n, to do the induction analysis. If we integrate on this last variable, and using the above inequality, we have:

$$V_n = \int_{-1}^{1} V_{n-1}\left(\sqrt{1 - x_n^2}\right) dx_n,$$

where $V_{n-1}(\sqrt{1 - x_n^2})$ is the cross-section of V_n at x_n. (Note that the cross-section of an n-dimensional hypersphere is always a $(n-1)$-dimensional hypersphere of radius $\sqrt{1 - x_n^2}$.)[2] So

$$V_n = \int_{-1}^{1} V_{n-1}\left(\sqrt{1 - x_n^2}\right) dx_n$$

$$= \int_{-1}^{1} \left(\sqrt{1 - x_n^2}\right)^{n-1} V_{n-1}(1)\, dx_n$$

$$= V_{n-1} \cdot \int_{-1}^{1} \left(1 - x_n^2\right)^{(n-1)/2} dx_n,$$

---

[2] For example a slice of a sphere ($n = 3$) at $z = x_3$ is a circle ($n = 2$) with radius $\sqrt{1 - x_3^2}$.

and we have:

$$V_2 = \pi, \quad V_4 = \frac{\pi^2}{2}, \quad V_6 = \frac{\pi^3}{6}, \dots, \quad V_{2n} = \frac{\pi^n}{n!}$$

$$V_3 = \frac{4\pi}{3}, \quad V_5 = \frac{8\pi^2}{15}, \quad V_7 = \frac{16\pi^3}{105}, \dots, \quad V_{2n+1} = \frac{\pi^n}{\frac{1}{2} \times \frac{3}{2} \times \cdots \times \frac{2n+1}{2}}.$$

The general formula for V_n is given by:

$$V_n = \frac{\pi^{n/2}}{\Gamma(n/2 + 1)} = \frac{\pi^{n/2}}{(n/2)!}$$

where Γ is the gamma function and $\Gamma(n + 1) = n!$ For the ratio follows:

$$\frac{V_{\text{hypersphere}}}{V_{\text{hypercube}}} = \frac{\pi^{n/2}}{(n/2)! 2^n} = \left(\frac{\pi}{4}\right)^{n/2} \cdot \frac{1}{(n/2)!}.$$

Since $\frac{\pi}{4} < 1$, we see that $V_n \to 0$ for $n \to \infty$. So the ratio $V_{\text{hypersphere}}/V_{\text{hypercube}}$ gets worse with increasing n, and this means that the above algorithm for the generation of equally distributed random directions is useless for large n. For example, the ratio for $n = 10$ is already $V_{10}/2^{10} \approx 0.0025$, i.e. you would expect to succeed only once every 400 trials.

An efficient generation of random directions is obtained by generating $d_i \in N(0, 1)$, a normally distributed number with average zero and variance one, for $i = 1, \dots, n$ and normalizing d by $d/\|d\|$. The only difficulty is to have a good generator for a $N(0, 1)$ distribution. For more details see Knuth, section 3.4.1.E(5), listed in the Further reading section.

A1.3.4 Newton's method of function minimization

Newton's famous iteration to approximate the solution of an equation can be applied to function minimization. This is done by approximating the gradient of f to zero for multivariate functions. We approximate the derivative of the function with a Taylor series and get:

$$f'(\mathbf{x} + \mathbf{\Delta x}) = f'(\mathbf{x}) + f''(\mathbf{x}) \cdot \mathbf{\Delta x}^{\text{T}} + \mathcal{O}\left(\|\mathbf{\Delta x}\|^2\right),$$

where $\mathbf{x} := (x_1, x_2, \dots, x_n)$ and $\mathbf{\Delta x} := (\Delta x_1, \Delta x_2, \dots, \Delta x_n)$.

Assuming that we are sufficiently close to the minimum, then $\|\mathbf{\Delta x}\|$ will be small and we ignore the $\mathcal{O}(\|\mathbf{\Delta x}\|^2)$ term. We find the $\mathbf{\Delta x}$ which will make all the derivatives zero on the assumption that the derivatives at the minimum are zero. That is

$$0 = f'(\mathbf{x} + \mathbf{\Delta x}) \approx f'(\mathbf{x}) + f''(\mathbf{x}) \cdot \mathbf{\Delta x}^{\mathrm{T}}$$

$$\text{or} \quad \mathbf{\Delta x}^{\mathrm{T}} \approx -f''(\mathbf{x})^{-1} \cdot f'(\mathbf{x}).$$

This requires the solution of a linear system of equations, where the matrix $f''(\mathbf{x})$ is the Hessian of $f(\mathbf{x})$ and the vector $f'(\mathbf{x})$ is the gradient of $f(\mathbf{x})$. Having computed the increment $\mathbf{\Delta x}$ we can add it to $\mathbf{x}$ to obtain a better approximation. In this way, by computing an increment for each vector $\mathbf{x}_i$, we obtain a sequence $\mathbf{x}_0, \mathbf{x}_1, \mathbf{x}_2, \ldots,$ where

$$\mathbf{x}_{i+1} = \mathbf{x}_i + \mathbf{\Delta x}_i$$

which should converge to a minimum.

To check that the method is working under our assumptions we will compute the following value:

$$c = \frac{f(\mathbf{x} + \mathbf{\Delta x}) - f(\mathbf{x}) - f'(\mathbf{x}) \cdot \mathbf{\Delta x}^{\mathrm{T}}}{\frac{1}{2}\mathbf{\Delta x} \cdot f''(\mathbf{x}) \cdot \mathbf{\Delta x}^{\mathrm{T}}}.$$

The computation of $f(\mathbf{x} + \mathbf{\Delta x})$ will be needed for the next step, so the computation of c requires only the additional calculation of a matrix-vector product. Under the assumption that $\|\mathbf{\Delta x}\|$ is sufficiently small, then c should be equal to 1. A reasonable check allowing for some small error (we ignored the $\mathcal{O}(\|\mathbf{\Delta x}\|^2)$ term), is to require for example

$$0.9 < c < 1.1.$$

Of course we should also check that $f(\mathbf{x} + \mathbf{\Delta x}) < f(\mathbf{x})$.

The assumption that we start sufficiently close to the minimum is a very important one, which has a serious consequence if it is not met: the $\mathcal{O}(\|\mathbf{\Delta x}\|^2)$ term is too significant to be ignored and hence we do not converge to a solution. The check for $0.9 < c < 1.1$ usually takes care of this problem.

To test whether we are converging to a minimum, maximum or saddle point, we have to test whether $f''(\mathbf{x})$ is positive definite. Since $f''(\mathbf{x})$ is a symmetric matrix, we can use Cholesky's decomposition to solve the linear system and, at the same time, check for $f''(\mathbf{x})$ to be positive definite. (Cholesky's method fails if the matrix is not positive definite.)

In case of failure, for any reason (Cholesky fails or c is out of range), we should use a different method: steepest descent or random directions if we are starting the minimization, or the spectral method if we are close to the end of the minimization. In the case that c is out of range we can also use Brent's algorithm in the direction given by $\mathbf{\Delta x}$.

Example For multivariate minimization:

$$f(\mathbf{x}) = \cos(x_1) + \frac{\sin(x_1)}{1 + x_2^2} \qquad f'(\mathbf{x}) = \begin{pmatrix} -\sin x_1 + \frac{\cos x_1}{x_2^2+1} \\ -2x_2 \frac{\sin x_1}{(x_2^2+1)^2} \end{pmatrix}$$

$$f''(\mathbf{x}) = \begin{pmatrix} -\cos x_1 - \frac{\sin x_1}{x_2^2+1} & -2x_2 \frac{\cos x_1}{(x_2^2+1)^2} \\ -2x_2 \frac{\cos x_1}{(x_2^2+1)^2} & -2\frac{\sin x_1}{(x_2^2+1)^2} + 8x_2^2 \frac{\sin x_1}{(x_2^2+1)^3} \end{pmatrix}.$$

With starting values $x_1 = -2.5$ and $x_2 = 0.1$ we get the following results.

Iteration	$\mathbf{x}$	$f(\mathbf{x})$	$f'(\mathbf{x})$	$f''(\mathbf{x})$		$\Delta \mathbf{x}$	$\|\Delta \mathbf{x}\|$	c
	x_1		f_{x_1}	$f_{x_1 x_1}$	$f_{x_1 x_2}$	Δx_1		
	x_2		f_{x_2}	$f_{x_2 x_1}$	$f_{x_2 x_2}$	Δx_2		
0	-2.5	-1.394	-0.195	1.394	0.157	0.154	0.1986	1.106
	0.1		0.117	0.157	1.127	-0.126		
1	-2.346	-1.4137	0.015	1.4137	-0.0358	-9.7×10^{-2}	0.027	0.993
	-0.026		-0.365	-0.0358	1.4228	2.5×10^{-2}		
2	-2.3558	-1.4142	4.5×10^{-4}	1.4142	-2.5×10^{-4}	3.1×10^{-4}	3.6×10^{-4}	0.99999
	-1.78×10^{-4}		-2.5×10^{-4}	-2.5×10^{-4}	1.4146	1.8×10^{-4}		
3	-2.3562	-1.4142	2.2×10^{-8}	1.4142	-7.9×10^{-8}	-1.6×10^{-8}	5.8×10^{-8}	
	-0.56×10^{-8}		-7.9×10^{-8}	-7.9×10^{-8}	1.4142	5.6×10^{-8}		

This has been computed to 20 digits of accuracy but only a few digits are shown.

A1.3.5 The spectral method for function minimization

The spectral method (SM) is a minimization iteration, which has as main goal avoiding saddle points. A saddle point is a point where the function has zero gradient, a minimum in some directions and a maximum in other directions. At a saddle point the gradient is zero, so methods like Newton cannot differentiate between a minimum, a maximum or a saddle point. Often Newton converges to saddle points or maxima instead of converging to a minimum. The spectral method addresses exactly this problem. If anything, the biggest difference between Newton and SM is that SM will avoid saddle points. Otherwise, both methods have quadratic convergence.

When we are using Newton with the Cholesky decomposition, we can detect that we are going into a saddle point, but we cannot do anything about it. The spectral method will avoid saddle points and move towards minima. For high-dimensional minimization, converging to saddle points is a real problem. Even for a dimension as low as $n = 10$ we expect $2^{10} - 2 = 1022$ saddle points for every maximum and every minimum. This is confirmed by our experience with real problems. So choosing a random starting point and searching for a zero of the gradient is very likely to take us to saddle points.

Mathematical derivation We approximate the function with a Taylor series as before:

$$f(\mathbf{x} + \Delta\mathbf{x}) = f(\mathbf{x}) + \Delta\mathbf{x} \cdot f'(\mathbf{x}) + \frac{1}{2}\Delta\mathbf{x} \cdot f''(\mathbf{x}) \cdot \Delta\mathbf{x}^{\mathrm{T}} + \mathcal{O}(\|\Delta\mathbf{x}\|^3),$$

where

$$\mathbf{x} := (x_1, x_2, \ldots, x_n), \quad \Delta\mathbf{x} := (\Delta x_1, \Delta x_2, \ldots, \Delta x_n), \ f'(\mathbf{x}) = \begin{pmatrix} f_{x_1}(\mathbf{x}) \\ f_{x_2}(\mathbf{x}) \\ \vdots \\ f_{x_n}(\mathbf{x}) \end{pmatrix},$$

$$f''(\mathbf{x}) = \begin{pmatrix} f_{x_1 x_1}(\mathbf{x}) & \cdots & f_{x_1 x_n}(\mathbf{x}) \\ \vdots & \ddots & \vdots \\ f_{x_n x_1}(\mathbf{x}) & \cdots & f_{x_n x_n}(\mathbf{x}) \end{pmatrix}.$$

We do an eigenvalue/eigenvector decomposition of the symmetric matrix $f''(\mathbf{x})$, the Hessian of the function, resulting in $f''(\mathbf{x}) = \mathbf{U}\boldsymbol{\Lambda}\mathbf{U}^{\mathrm{T}}$ with

$$\boldsymbol{\Lambda} = \begin{pmatrix} \lambda_1 & & 0 \\ & \ddots & \\ 0 & & \lambda_n \end{pmatrix},$$

where the λ_i are the eigenvalues of $f''(x)$ and $\mathbf{U}$ is the matrix of the eigenvectors of $f''(x)$ and where $\mathbf{U}\mathbf{U}^{\mathrm{T}} = \mathbf{1}$.

We set $\mathbf{z} = \Delta\mathbf{x} \cdot \mathbf{U}$ and $\mathbf{d} = \mathbf{U}^{\mathrm{T}} \cdot f'(\mathbf{x})$. With these substitutions we can write the above equation as:

$$f(\mathbf{x} + \Delta\mathbf{x}) = f(\mathbf{x}) + \underbrace{\Delta\mathbf{x} \cdot \mathbf{U}}_{\mathbf{z}} \underbrace{\mathbf{U}^{\mathrm{T}} \cdot f'(\mathbf{x})}_{\mathbf{d}} + \frac{1}{2} \underbrace{\Delta\mathbf{x} \cdot \mathbf{U}}_{\mathbf{z}} \boldsymbol{\Lambda} \underbrace{\mathbf{U}^{\mathrm{T}} \cdot \Delta\mathbf{x}^{\mathrm{T}}}_{\mathbf{z}^{\mathrm{T}}} + \mathcal{O}(\|\Delta\mathbf{x}\|^3)$$

$$= f(\mathbf{x}) + \mathbf{z} \cdot \mathbf{d} + \frac{1}{2}\mathbf{z} \cdot \boldsymbol{\Lambda} \cdot \mathbf{z}^{\mathrm{T}} + \mathcal{O}(\|\Delta\mathbf{x}\|^3).$$

So

$$f(\mathbf{x} + \Delta\mathbf{x}) - f(\mathbf{x}) \approx \underbrace{\mathbf{z} \cdot \mathbf{d}}_{\text{scalar product}} + \frac{1}{2} \underbrace{\mathbf{z} \cdot \mathbf{\Lambda} \cdot \mathbf{z}^{\mathrm{T}}}_{\text{vector} \cdot \text{matrix} \cdot \text{vector}}$$

$$\approx \sum_{i=1}^{n} z_i d_i + \frac{1}{2} z_i^2 \lambda_i,$$

since $\mathbf{\Lambda}$ is a diagonal matrix. The change of $f(\mathbf{x})$ near $\mathbf{x}$ depends on the z_i, d_i and λ_i. The d_i and λ_i are determined by f and x, but we can choose the z_i as we like.

The $\sum_{i=1}^{n} z_i d_i + \frac{1}{2} z_i^2 \lambda_i$ is just a sum of n quadratic equations in the z_i, i.e. n parabolas. We can choose the z_i independently of each other to minimize each term, so we can minimize $g(z_i) = z_i d_i + \frac{1}{2} z_i^2 \lambda_i = z_i(d_i + \frac{1}{2} z_i \lambda_i)$ independently for each i.

Let us look at the different possibilities.

(i) $\lambda_i > 0$, i.e. we have a parabola which is open at the upper end:
    (a) $d_i < 0$ (see Figure A1.4),
    (b) $d_i > 0$, identical calculation to the case above (see Figure A1.5),
    (c) $d_i = 0$ (see Figure A1.6).

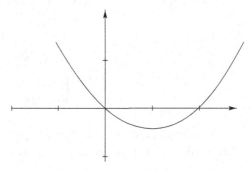

Figure A1.4   The situation for $\lambda_i > 0$ and $d_i < 0$. The minimum occurs at $g'(z_i) = d_i + z_i \lambda_i = 0$ or $z_i = -d_i/\lambda_i > 0$.

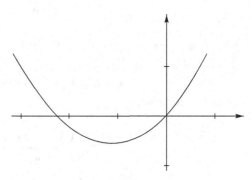

Figure A1.5   The situation for $\lambda_i > 0$ and $d_i > 0$. The minimum occurs for $z_i = -d_i/\lambda_i < 0$.

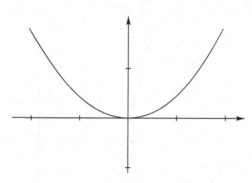

Figure A1.6   The situation for $\lambda_i > 0$ and $d_i = 0$. The minimum occurs for $z_i = 0$.

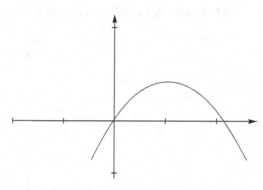

Figure A1.7   The situation for $\lambda_i < 0$ and e.g. $d_i > 0$. We have a maximum at $z_i = -d_i/\lambda_i$.

(ii) $\lambda_i < 0$, i.e. we have a parabola, which is open at the lower end (see Figure A1.7). Since it is open at the lower end, it does not have a minimum. The parabola values will become arbitrarily negative by choosing $|z_i|$ large enough.

If we use Newton, it would lead us to the maximum, i.e. in the wrong direction. So in this case we have to go in the opposite direction. Since the maximum is at $z_i = -d_i/\lambda_i$, we have to move into the direction of $-\text{sign}(-d_i/\lambda_i)$ to minimize the function. But we do not know how far we have to go. (Notice that as we move farther from $z_i = 0$, our $\|\Delta x\|$ becomes larger and our approximation becomes worse.) Let us call this unknown distance h, $h \geq 0$. Then our minimum would be at $z_i = -\text{sign}(-d_i/\lambda_i)h = -\text{sign}(d_i)h$, since λ_i is negative.

The basis for all our calculations was an approximation. We developed the function around x up to second order and ignored the $\mathcal{O}(\|\Delta x\|^3)$ terms. This approximation only holds in a neighborhood of x. So this does not mean, that we do not have another minimum (see Figure A1.8).

We choose the direction where we minimize the function with the closest z_i, i.e. smallest $\|\Delta x\|$. This is to use the best possible approximation.

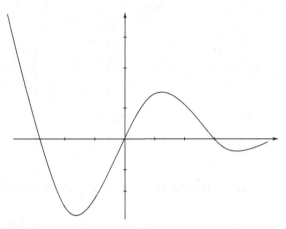

Figure A1.8   A picture showing a possible global situation.

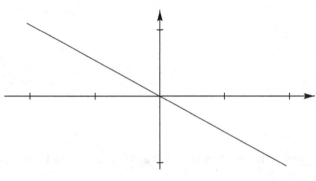

Figure A1.9 $\lambda_i = 0$ and $d_i < 0$.

    In the case where $d_i = 0$ (we are at the summit of the parabola) we can choose any direction with equal chance of success. Say, we choose "right," i.e. if $\lambda_i < 0$, $d_i = 0$ then $z_i = h$.

(iii) When $\lambda_i = 0$, there are again three possible cases:

    (a) $d_i < 0$ (see Figure A1.9),

    (b) $d_i > 0$ (see Figure A1.10), in both cases the minimum is at $z_i = -\text{sign}(d_i) \cdot h$ (h is unknown),

    (c) $d_i = 0$ (see Figure A1.11). In this case we have approximated the curve with a straight horizontal line. Either we are at a peculiar point of the curve or this direction (dimension) does not matter at all for the minimum. The function is insensitive to this variable (at least to second order). So the best choice is to do nothing and set $z_i = 0$.

By doing this analysis for each $i \in \{1, \ldots, n\}$ we get a value for each z_i. So now we know the vector $\mathbf{z} = (z_1, z_2, \ldots, z_n)$, for example $(1.2, -h, 0.5, \ldots, h, \ldots)$ and

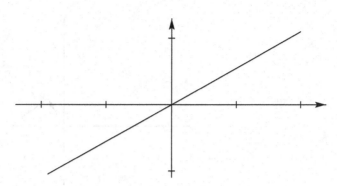

Figure A1.10 $\lambda_i = 0$ and $d_i > 0$.

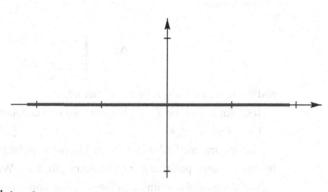

Figure A1.11 $\lambda_i = 0$ and $d_i = 0$.

$\Delta x = zU^T$ will contain only numbers and the symbol h. This means that Δx is linear in h, something like $(1.8, 1.7h + 0.1, \ldots)$.

We can write $\Delta x := \Delta x_0 + \Delta x_1 \cdot h$, where Δx_0 is the part containing only numbers and Δx_1 are the factors of h (which could be zero). If $\Delta x_1 \neq 0$ we have to choose some $h > 0$ to minimize the functional. This is nothing other than choosing a minimum in the direction of Δx_1 after having moved to Δx_0 (Figure A1.12).

A variant of the above algorithm, which is simpler to code, is to use the ideas behind steepest descent for the cases where $\lambda_i < 0$. In those cases we are searching in a particular direction, but we do not know how far to search (we use the parameter h to determine later, how far to search). Using the ideas of steepest descent (see Section A1.3.2) we should make our direction of search proportional to the partial derivatives. In our case, when $\lambda_i < 0$ the derivative at the point is d_i. Hence our increment for z_i should be $-d_i h$. (Recall that in the previous assignment we used $-\text{sign}(d_i)h$ and 0 when $d_i = 0$.)

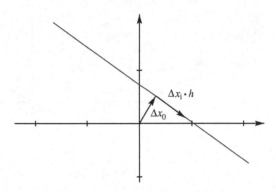

Figure A1.12 Minimizing in the direction of x_1 after having moved to x_0.

In summary

$$\Delta z_i = \begin{cases} -\dfrac{d_i}{\lambda_i} & \text{if } \lambda_i > 0 \\ -d_i h & \text{if } \lambda_i \leq 0 \end{cases}$$

and the rest of the algorithm is as before.

Tests for $\lambda_i = 0$ or $d_i = 0$ should take into account the roundoff error, i.e. test for $|\lambda_i| \leq \varepsilon$ and $|d_i| \leq \varepsilon$.

The spectral method solves the saddle point problem: a saddle point identifies itself by having some positive λ_i and some negative λ_j. With Newton we minimize in the i-dimensions and maximize in the j-dimensions. The spectral method identifies the j-dimensions and chooses the right direction (away from the maximum) in every case.

Please see the interactive exercise "Spectral method."

A1.4 Comparison of minimization methods

A1.4.1 The amount of computation needed in each step of the approximation

Random directions Random directions (RD) requires computing a random direction and then doing Brent iterations in this particular direction. Each iteration involves only the computation of the value of the function, and we do this as many times as we do Brent steps. So we compute only one value of the function and the complexity is $T_f \times B$, where B is the number of Brent iterations, and T_f the time to compute one value of $f(x)$.

Steepest descent Steepest descent (SD) computes the direction of steepest descent which requires the gradient. Hence it requires that we compute n partial derivatives. We cannot be very precise about how much it takes to compute the derivatives: it may take less than the computation of the functional value $f(x)$ or it may take more. So we will assume that each derivative has a cost equivalent to the functional. Since we compute the function and n derivatives the amount of time needed is $(n + 1) \cdot T_f$. With the gradient we find the direction of steepest descent and then we run the Brent algorithm, this gives us a total complexity of $n + 1 + B$ function evaluations per step.

Newton Newton computes the functional, the first derivatives (the gradient) and the second derivatives (the Hessian). So the complexity of the function evaluation part is $(n + 2)(n + 1)/2 \cdot T_f$, since the Hessian is a symmetric matrix and needs $n(n + 1)/2$ entries computed. We select a direction, and then we use Brent, which means B steps. To solve the system of n linear equations, we use either gaussian elimination or Cholesky decomposition (as is recommended) and both are $\mathcal{O}(n^3)$ problems. So we have $\mathcal{O}(n^3)$ matrix operations and $(n + 2)(n + 1)/2 + B$ function evaluations per step.

Spectral method The spectral method is complexity-wise very similar to Newton. We need to compute the first and second derivatives, we end up finding a direction in which we use Brent, and we solve an eigenvector/eigenvalue problem. This method requires $\mathcal{O}(n^3)$ matrix operations and $(n + 2)(n + 1)/2 + B$ for the function evaluations per step.

Both Newton and the spectral method require $\mathcal{O}(n^3)$ matrix operations. In practice gaussian elimination or Cholesky's decomposition are much faster than eigenvalue/eigenvector decomposition, about a factor of 10 faster.

So, we can summarize the complexity of the multidimensional minimization methods as follows.

Method	Function evaluations	Matrix operations	Convergence
Random directions	B		linear
Steepest descent	$(n + 1 + B)$		linear
Newton	$\left(\frac{(n+2)(n+1)}{2} + B \right)$	$\mathcal{O}(n^3)$	quadratic
Spectral method	$\left(\frac{(n+2)(n+1)}{2} + B \right)$	$\mathcal{O}(n^3)$	quadratic

Conclusions

(PRACTICAL NOTE)

- Random directions (RD) is the cheapest method per iteration. So we want to use it as often as it is useful.

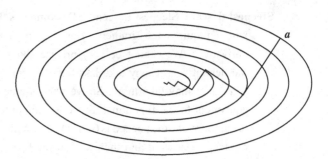

Figure A1.13   A steepest descent path starting from point a and converging to the minimum in about eight iterations. Note how consecutive steps consistently overshoot the best path to the minimum, resulting in an (inefficient) stair-like trajectory.

- RD does not require the computation of derivatives which for some problems may be a huge advantage: in some cases derivatives may not exist, in other cases the formulas may be too complicated to code them (and correctness is a serious problem).

- RD is *very* effective at the beginning, since we will practically always find a direction which is far from horizontal. This ceases to be the case when we are close to the minimum. Close to the minimum RD will be very close to horizontal lines in terms of $f(x)$.

- Steepest descent (SD) is the second choice for the start of the minimization process. It is also inexpensive compared to Newton and the spectral method.

- RD and SD share a lot of properties: both are very slow close to the end of the minimization process, i.e. near the minimum, both are good at the start, and both are relatively inexpensive.

- SD has a stair-like behavior (see Figure A1.13): near the minimum the directions of consecutive steps will be at nearly 90° angles. This slows down the minimization.

- Newton (N) and the spectral method (SM) have quadratic convergence, i.e. they are very good when close to the minimum. If the order of convergence is ρ, then

$$\|x_{n+1} - x_n\| = \mathcal{O}\left(\|x_n - x_{n-1}\|^{\rho}\right),$$

so with quadratic convergence ($\rho = 2$), the number of accurate digits is basically doubled for each step.

- RD and SD are cheap and simple – they are very good at the beginning. N and SM are more expensive, but very good towards the end.

- Newton should always be used with protection against saddle points and maxima, i.e. using Cholesky decomposition. This gives at least a signal: if there is a saddle point, the decomposition cannot be computed. Cholesky's algorithm decomposes a matrix $\mathbf{A}$, appearing in an equation $\mathbf{A}x = b$ as $\mathbf{A} = \mathbf{R}\mathbf{R}^{\mathrm{T}}$, where $\mathbf{R}$ is lower triangular. If **A** is not positive definite, then **R** cannot be computed. The Newton step cannot be computed, and some other method has to be used (e.g. SD), but we are prevented from converging to saddle points or maxima. If we use Newton without Cholesky, then Newton will converge almost inevitably to saddle points instead of minima for high dimensions, e.g. $n > 10$.

- When Newton fails due to a non-positive-definite **A**, we can either use SD (at the beginning) or SM (at the end).
- SM is the safest but most expensive algorithm per iteration.
- In the case that $f(x)$ is actually dependent on fewer than n indeterminates, N will be forced to solve singular or (because of rounding errors) nearly singular problems and will usually fail. SM is robust in this sense.

Recommended procedure Use RD (maybe SD) at the beginning. When we detect that we do not make much progress, switch to N with Cholesky, and whenever N fails use SM.

So we win speed at the beginning and get close to a local minimum, so that N has less chance of failing, and we win in the end with the quadratic methods.

Why should we code Newton *and* the spectral method? Why not code SM only which has quadratic convergence and is safe? Eigenvalue/eigenvector solutions are about 10 times more expensive than Cholesky's decomposition. So when we have a very large problem with $n > 1000$, we have to minimize the number of SM iterations and use N as often as possible.

A1.4.2 Final comments

Why do we need continuous functions with continuous derivatives?

The answer is quite obvious: almost all methods rely on some kind of Taylor expansion. In N and SM the constants in the $\mathcal{O}(\|\Delta\mathbf{x}\|^3)$ term depend on the third derivatives. So if a third derivative happens to be infinite at some point, we cannot bound the errors.

So discontinuities in f, f' or f'' will invalidate the approximation and we may not have quadratic convergence. Indeed often we will have **no convergence at all**!

When do such discontinuities arise? Such discontinuities may be inherent to the problem, but sometimes they are introduced by oversights.

(PRACTICAL NOTE)   **Simple example** Suppose we have a molecular dynamics problem, where we have a structure with three atoms, C, O and S. Suppose the C and O are close together but the S is far away. When the C and the O are close, the van der Waals forces are

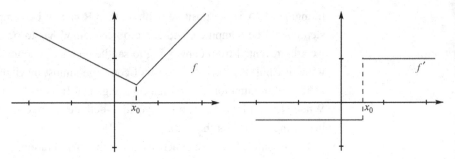

Figure A1.14 **A function of the type $f(x) = |x - x_0| + \frac{x}{a} + b$ and its discontinuous derivative.**

relevant. For the very distant S the van der Waals forces are irrelevant, since they are insignificantly small. They are completely negligible since they are proportional to $\mathcal{O}(1/d^6)$. Normally we only compute the van der Waals force inside a circle around the atom group. Now we have a boundary: inside we compute with the van der Waals forces, outside we neglect them. And however small the van der Waals forces are at the boundary, we are creating a discontinuity, and this discontinuity may invalidate our approximation.

So it is important to pay attention and make the functions really continuous and differentiable, and avoid introducing artificial discontinuities (in the first derivative) by trying to save computing time. Only in the case that the contribution of a force will fall out of precision (and hence it makes absolutely no difference to the computed result) can we ignore a contribution. Discontinuities also arise with the use of absolute value; in the case shown in Figure A1.14 we have a discontinuous first derivative.

Computational precision

Should we use single precision, double precision or quadruple precision?

When you look at a problem in energy minimization you may find, for example, that the distance between a carbon and a nitrogen atom is $d(C, N) = 1.8 \pm 0.05$ Å.

For most stock market simulations, seven digits of precision mean less than a dollar for **all** the transactions of a day, clearly insignificant. Stock prices are rounded to the nearest cent, which means a relative error of about one in 20 000.

Single precision gives us about seven digits of precision, so should we even think about using more than single precision, when the initial data have merely two significant digits?

This question is seductive, but wrongly posed!

When we say that the separation between the two atoms is 1.8 Å, then we have a big error only because we do not know this value with higher precision.

But in our modelling, these two atoms and all other carbon and nitrogen atoms are separated by exactly the same distance. So for the minimization process this value of 1.8 Å should be considered exact.

The minimization process is very sensitive to numerical errors, because approximation errors cause discontinuities. If you are working with seven digits, then some discontinuities will be introduced, which will imply that our approximations may become invalid.

So the precision that we have to use is not related to the precision of the data, it is not even related to the precision of our results, it is related to the precision needed to run the minimization process.

The precision is needed for the minimization algorithm to be successful.

This is an important point, which is quite often overlooked, because of the argument: "Oh, your data are so bad, just use 2 digits." No, computing with two digits of precision you will not go anywhere, because you will never be able to run the minimization algorithms successfully. Even single precision may not be enough. As a matter of fact you will not be able to invert successfully a matrix of 1000×1000 or compute an eigenvalue/eigenvector decomposition in single precision – the errors will normally be too large. Usually double precision is needed.

A1.5 Confidence limits of sums and averages

Confidence limits of sums and averages.

Let x be a random variable with finite expected value $E[x]$ and finite variance $\sigma^2(x)$.

An average of $x$es

$$\bar{x} = \frac{\sum_{i=1}^{n} x_i}{n}$$

or a sum of $x$es is normally distributed as $n \to \infty$. This allows us to conclude confidence limits of the estimates of sums and averages.

$$\Pr\{|\bar{x} - E[x]| > 1.96\sigma(x)\} = 0.05.$$

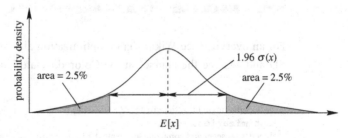

Figure A1.15   Probabilities and multiples of the standard deviaton

The constants multiplying the standard deviation are

$$1.959963\ldots \quad \text{for } p = 1/20$$
$$2.575829\ldots \quad \text{for } p = 1/100$$
$$3.290526\ldots \quad \text{for } p = 1/1000$$

The variance of x is estimated by

$$\sigma^2(x) \approx \frac{n \sum x_i^2 - \left(\sum x_i\right)^2}{n(n-1)}.$$

Acknowledgements

The authors want to acknowledge and thank the help and suggestions provided by: Gina Cannarrozi, Daniel Dalquen, Christophe Dessimoz, Manuel Gil, Adrian Schneider, Adam Szalkowski and Adreas Vogt.

FURTHER READING

Sections A1.2–A1.3.2 are based on the following.

W. H. Press, B. P. Flannery, S. A. Teukolsky and W. T. Vetterling, *Numerical Recipes in C: the Art of Scientific Computing*, Cambridge University Press, 1992, Chapter 10.

R. P. Brent, *Algorithms for Minimization without Derivatives*, Englewood Cliffs, NJ: Prentice-Hall, 1973.

D. E. Knuth, *The Art of Computer Programming*, Volume 2, *Seminumerical Algorithms*, Addison-Wesley, 1969.

Online material for minimization methods

For an overview see Wikipedia on optimization algorithms.[3] For an introduction to SD and CG see the course[4] at the Colorado State University. Material on Brent's

---

[3] Wikipedia on optimization: `http://en.wikipedia.org/wiki/Category:Optimization_algorithms`

[4] Steepest descent and conjugate gradients: `http://cauchy.math.colostate.edu/Resources/SD_CG/index.html`

method,[5] Cholesky factorization[6] and steepest descent and conjugate gradients[7] can be found in this undergraduate course on numerical analysis.[8]

Online material for linear least squares

There is a nice applet displaying the errors when approximating data points by a straight line, i.e. linear least squares.[9] (The user may choose an approximation – the errors are computed in real time. The user may also define the points, the real solution is computed.) There is another applet,[10] where the user can insert, move and delete points, and the minimizing line is calculated and shown in real time, and a third nice applet,[11] where the user can try to find the minimizing line. (The errors are calculated and shown in real time.)

Online material for non-linear least squares

An introduction to non-linear least squares (including some other methods like Gauss–Newton and the Levenberg–Marquardt) can be found at the Argonne National Laboratory.[12]

---

[5] Brent: http://math.fullerton.edu/mathews/n2003/BrentMethodMod.html
[6] Cholesky: http://math.fullerton.edu/mathews/n2003/CholeskyMod.html
[7] SD and CG: http://math.fullerton.edu/mathews/n2003/GradientSearchMod.html
[8] Numerical analysis: http://math.fullerton.edu/mathews/n2003/NumericalUndergradMod.html
[9] Linear least squares applet 1: www.ies.co.jp/math/java/misc/least_sq/least_sq.html
[10] Linear least squares applet 2: www.barrodale.com/java_demo/javademo.htm
[11] Linear least squares applet 3: http://mathforum.org/dynamic/java_gsp/squares.html
[12] Non-linear LS: www-fp.mcs.anl.gov/otc/Guide/OptWeb/continuous/unconstrained/nonlinearls/

Appendix B **Online resources**

B1.1 General online material

Base of scientific computing related information
http://openlink.br.inter.net/sparse/sci-comp.html

Scientific computing links
www.phy.duke.edu/~hsg/sci-computing.html

Scientific computing and associated fields resource guide
www.mathcom.com/corpdir/techinfo.mdir/scifaq/index.html

SIAM Journal of Scientific Computing
http://epubs.siam.org/sam-bin/dbq/toclist/SISC

The German scientific computing initiative homepage (in English)
www.scicomp.uni-erlangen.de/index2.html

B1.2 Institutes for scientific computing at universities

(A very incomplete list, provided as a starting point for students.)
ETH-Zürich, Institute for Computational Science (ICOS)
www.icos.ethz.ch/

Uni Bonn, Institut für numerische Simulation
wwwwissrech.iam.uni-bonn.de/

TU München, Wissenschaftliches Rechnen
https://www-m3.ma.tum.de/twiki/bin/view/Allgemeines/

Weierstrass-Institut Berlin, Forschungsgruppe Numerische Mathematik und Wissenschaftliches Rechnen
www.wias-berlin.de/research-groups/nummath/

TU Braunschweig, Institut für Wissenschaftliches Rechnen
www.tu-bs.de/institute/WiR/institutofr.html

Universität Heidelberg, Institut für Wissenschaftliches Rechnen
www.iwr.uni-heidelberg.de/

Boston University, Scientific Computing and Visualization Group
http://scv.bu.edu/

Stanford University, Institute for Computational and Mathematical Engineering
www-sccm.stanford.edu/

North Carolina State University, Center for Research in Scientific Computation
(CRSC)
www.ncsu.edu/crsc/

University of Maryland, Center for Scientific Computation and Mathematical
Modeling (CSCAMM)
www.cscamm.umd.edu/

University of Utah, Scientific Computing and Imaging Institute
www.sci.utah.edu/

Rensselaer Polytechnic Institute, Scientific Computation Research Center (SCOREC)
www.scorec.rpi.edu/

Warwick University, Centre for Scientific Computing
www2.warwick.ac.uk/fac/sci/csc/

B1.3 Scientific computing at other research facilities

International Max Planck Research School for Computational Biology and Scientific
Computing (IMPRS-CBSC)
www.mpg.de/instituteProjekteEinrichtungen/schoolauswahl/computeBiology/

Alfred-Wegener-Institut für Polar- und Meeresforschung Bremerhaven, Scientific
Computing Group
www.awi-bremerhaven.de/InfoCenter/IT/WorkingGroups/SciComp/

Zuse-Institut Berlin, Scientific Computing Division
www.zib.de/General/Divisions/SC/index.en.html

Lawrence Livermore National Laboratory, Center for Applied Scientific Computing
www.llnl.gov/CASC/

Pittsburgh Supercomputing Center, Projects in Scientific Computing
www.psc.edu/science/

Lawrence Berkeley National Laboratory, National Energy Research Scientific Computing Center
www.nersc.gov/

National Center for Atmospheric Research in Boulder, Colorado, Scientific Computing Division
www.scd.ucar.edu/main/info.html

B1.4 Mathematical software resources

Sources of mathematical software
www.cse.uiuc.edu/heath/scicomp/software.html

Handbook of algorithms and data structures (containing the essential algorithms in Pascal and C code)
http://sunsite.dcc.uchile.cl/~rbaeza/handbook/expand.html

The numerical recipes homepage
www.nr.com/

Index

Printed in the United States
by Baker & Taylor Publisher Services